Telecommunications: An Engineering Perspective

Telecommunications: An Engineering Perspective

Claire Adams

MURPHY & MOORE
www.murphy-moorepublishing.com

Published by Murphy & Moore Publishing,
1 Rockefeller Plaza,
New York City, NY 10020, USA

ISBN: 978-1-63987-518-4

Cataloging-in-Publication Data

Telecommunications : an engineering perspective / Claire Adams.
    p. cm.
Includes bibliographical references and index.
ISBN 978-1-63987-518-4
1. Telecommunication. 2. Telecommuting. 3. Telecommunication systems.
4. Engineering. 5. Telecommunications engineers. I. Adams, Claire.
TK5101 .T45 2022
621.382--dc23

For information on all Murphy & Moore Publications
visit our website at www.murphy-moorepublishing.com

 MURPHY & MOORE

Contents

Preface

The process by which information such as messages, images, signals and sounds are transferred over distances using wire, radio, optical or electromagnetic systems is referred to as telecommunications. These systems or communication paths are often divided into multiple channels to facilitate multiplexing. Telecommunication technologies are divided into two major branches, wired and wireless. Some examples of telecommunications systems are internet, radio, telephone, etc. The three basic elements of any telecommunications system are transmitter, transmission medium and receiver. Telecommunications engineering focuses on designing and maintaining such systems. This book attempts to understand the multiple branches that fall under the discipline of telecommunications engineering and how such concepts have practical applications. Some of the diverse topics covered herein address the varied branches that fall under this category. This textbook is an essential guide for both academicians and those who wish to pursue this discipline further.

A detailed account of the significant topics covered in this book is provided below:

Chapter 1- Telecommunication can be defined as the process of information transmission through voice, data or video through electromagnetic signals. All types of telecommunication employ the same basic elements consisting of a transmitter, transmission medium and a receiver to operate. The physical or logical medium used for the transmission is called a communication channel. This chapter's aim is to introduce the reader to telecommunication.

Chapter 2- A telecommunications network is a transmission system consisting of a collection of interconnected nodes, links and intermediate nodes that allow the transmission of information. There are many types of telecommunications networks employed globally like the wide area networks, local area networks, virtual private networks, client/server networks, etc. This chapter has been carefully written to provide the reader with a better understanding of the subject matter.

Chapter 3- Wireless technology is the technology that allows communication to take place without the need for connecting wires or cables. It is responsible for the creation and widespread adoption of wireless network. Wireless LAN, wireless access point, wireless WAN, wireless mesh network, etc. are all examples of modern wireless networks. This chapter is a comprehensive summary of modern wireless networks, its affiliated concepts and applications.

Chapter 4- Multiplexing is a popular networking technique in which multiple analog and digital signals are combined into one signal over a shared medium. This allows for a large number of signals to be transmitted over one medium. Time-division multiplexing, frequency-division multiplexing, space-division multiplexing, polarization-multiplexing, orbital angular momentum multiplexing, wavelength-division multiplexing, etc. are all types of multiplexing which are thoroughly covered in this chapter.

Chapter 5- The process of converting data into an electronic or optical carrier signal and its subsequent transmission is known as modulation. The two types of modulation discussed in this chapter are analog and digital modulation. This chapter's aim is to shed light on the topic of modulation and all of its related components for the benefit of the reader.

It gives me an immense pleasure to thank our entire team for their efforts. Finally in the end, I would like to thank my family and colleagues who have been a great source of inspiration and support.

Claire Adams

An Introduction to Telecommunication

Telecommunication can be defined as the process of information transmission through voice, data or video through electromagnetic signals. All types of telecommunication employ the same basic elements consisting of a transmitter, transmission medium and a receiver to operate. The physical or logical medium used for the transmission is called a communication channel. This chapter's aim is to introduce the reader to telecommunication.

The term telecommunications generally refers to all types of long-distance communication that use common carriers, including telephone, television, and radio. Data communications is a subset of telecommunications and is achieved through the use of telecommunication technologies.

In modern organizations, communications technologies are integrated. Businesses are finding electronic communications essential for minimizing time and distance limitations. Telecommunications plays a special role when customers, suppliers, vendors, and regulators are part of a multinational organization in a world that is continuously awake and doing business somewhere 24 hours a day, 7 days a week ("24/7").

A telecommunications system is a collection of compatible hardware and software arranged to communicate information from one location to another. These systems can transmit text, data, graphics, voice, documents, or video information. Such systems have two sides: the transmitter and the receiver. The major components are:

- Hardware: All types of computers and communications processors (such as a modems or small computers dedicated solely to communications).

- Communications media: The physical media through which electronic signals are transferred; includes both wire line and wireless media.

- Communications networks: The linkages among computers and communications devices.

- Communications processors: Devices that perform specialized data communication functions; includes front-end processors, controllers, multiplexors, and modems.

- Communications software: Software that controls the telecommunications system and the entire transmission process.

- Data communications providers: Regulated utilities or private firms that provide data communications services.

- Communications protocols: The rules for transferring information across the system.

- Communications applications: Electronic data interchange (EDI), teleconferencing, videoconferencing, e-mail, facsimile, electronic funds transfer, and others.

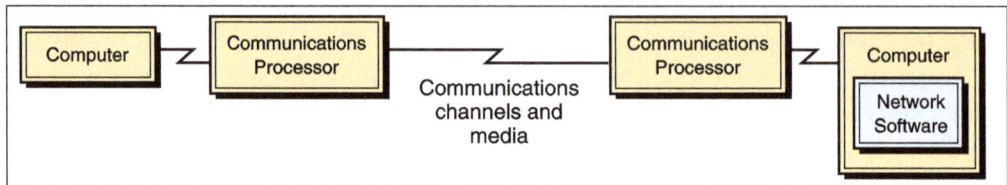

An integrated computer and telecommunications system.

To transmit and receive information, a telecommunications system must perform the following separate functions that are transparent to the user:

- Transmit information.

- Establish the interface between the sender and the receiver.

- Route messages along the most efficient paths.

- Process the information to ensure that the right message gets to the right receiver.

- Check the message for errors and rearrange the format if necessary.

- Convert messages from one speed to that of another communications line or from one format to another.

- Control the flow of information by routing messages, polling receivers, and maintaining information about the network.

- Secure the information at all times.

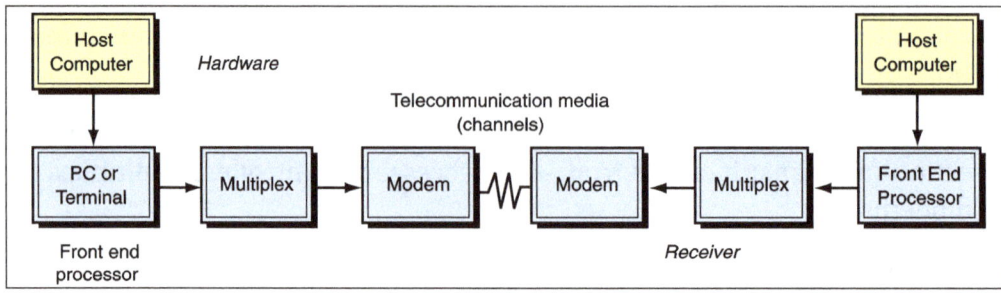

A telecommunications system.

Telecommunications media can carry two basic types of signals, analog and digital. Analog signals are continuous waves that "carry" information by altering the amplitude and frequency of the waves. For example, sound is analog and travels to our ears in the form of waves—the greater the height (amplitude) of the waves, the louder the sound; the more closely packed the waves (higher frequency), the higher the pitch. Radio, telephones, and recording equipment historically transmitted and received analog signals, but they are rapidly changing to digital signals. Digital signals are discrete on-off pulses that convey information in terms of 1's and 0's, just like the central processing unit in computers.

Digital signals have several advantages over analog signals. First, digital signals tend to be less affected by interference or "noise." Noise (e.g., "static") can seriously alter the information carrying characteristics of analog signals, whereas it is generally easier, in spite of noise, to distinguish between an "on" and an "off." Consequently, digital signals can be repeatedly strengthened over long distances, minimizing the effect of any noise. Second, because computer-based systems process digitally, digital communications among computers require no conversion from digital to analog to digital. Computer-based systems process digitally; digital communications among computers require no conversion from digital to analog to digital.

Communications processors are hardware devices that support data transmission and reception across a telecommunications system. These devices include modems, multiplexers, front-end processors, and concentrators.

A modem is a communications device that converts a computer's digital signals to analog signals before they are transmitted over standard telephone lines. The public telephone system (called POTS for "Plain Old Telephone Service") was designed as an analog network to carry voice signals or sounds in an analog wave format. In order for this type of circuit to carry digital information, that information must be converted into an analog wave pattern. The conversion from digital to analog is called modulation, and the reverse is demodulation. The device that performs these two processes is called a modem, a contraction of the terms modulate/ demodulate. Modems are always used in pairs.

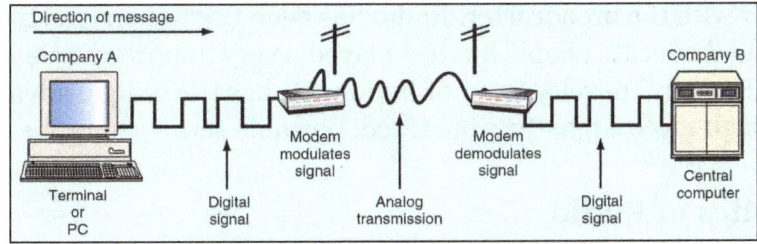

A modem converts digital to analog signals and vice versa.

Digital subscriber line (DSL) service allows the installed base of twisted-pair wiring in the telecommunications system to be used for high-volume data transmission. DSL uses digital transmission techniques over copper wires to connect the subscribers to network equipment located at the telephone company central office. Asymmetric DSL (ADSL) is a variety of DSL that enables a person connecting from home to upload data at speeds from 16 to 640 Kbps and download data at 1.5 to 8 Mbps. Clearly, this is many times faster than an analog modem.

A DSL circuit connects a DSL modem on each end of a twisted-pair telephone line, creating three information channels—a high-speed downstream channel; a medium-speed duplex channel, depending on the implementation of the DSL architecture; and POTS (Plain Old Telephone Service) or an ISDN channel. The POTS/ISDN channel is split off from the digital modem by filters, thus guaranteeing uninterrupted POTS/ISDN, even if DSL fails.

Cable modems are offered by cable television companies in many areas as a high-speed way to access a telecommunications network. These modems operate on one channel of the TV coaxial cable. Cost and transmission speed are comparable to that of a DSL. A cable modem gives users high-speed Internet access through a cable TV network at more than 1 Mbps (1 million bits per second), or about 20 times faster than a traditional dial-up modem. When a cable modem unit is installed next to the computer, a splitter is placed on the side of the household. It separates the coaxial cable line serving the cable modem from the line that serves the TV sets. A separate coaxial cable line is then run from the splitter to the cable modem. Cable modems typically connect to computers through a standard 10Base-T Ethernet interface. Data are transmitted between the cable modem and computer at 10 Mbps.

Types of Telecommunication

Radio

Radio is the most popular means of mass communication across the globe today. It is exclusively a medium of the sound. It is a medium where a performer cannot see his/her audience. And the audience too cannot see the performer. That is why it is often referred to as a "blind medium". There is ample space for imagination just as a book has. It is also an intimate medium, as the broadcaster always addresses the listener in singular form as if he/she is talking to him alone. Because of this, the listeners too feel a direct connect with the broadcaster. Radio has been used as an effective medium not only to inform and educate people but has played a very important role in the economic, political and cultural development of nations. It has the unique advantage of being receivable through low cost, battery operated, portable sets, even in the rural areas.

Characteristics of Radio

After print media, radio has emerged as a powerful and influential means of mass communication. It has successfully penetrated to every nook and corner of the country and been able to impact human culture. While television has come up in a big way, the popularity of radio still remains intact.

1. Radio is a medium of the voice: Similarly, if you listen to the commentary or a ball-by-ball account of a cricket match on radio, you are able to create the vivid visual imagery of a cricket stadium, the excitement on field and how the game is being played. The audio narration performs on the canvas of the listener's mind and the mind can then construct any period, any place. Radio is therefore exclusively a medium of the sound and can make pictures in your mind.

2. Radio is an intimate medium: The broadcaster always addresses the listeners as if he is talking to him/her alone and tries to build an instant rapport. The listeners too feel connected with him, if he addresses issues close to their heart. Every broadcaster knows that if he listeners do not like his programme, they will always have the option of switching channels. Therefore, any programme—be it a talk show, a documentary or a feature—must strike a chord with the listener, right at the beginning.

3. Radio as a mass medium: Though radio started as a communication tool for the armed forces, it soon became popular among the masses. The best part about radio is that it reaches millions of people at the same time. The audience may comprise people from different educational, social and cultural background. The broadcaster's job is to find out the lowest common denominator to communicate well with maximum number of people. Radio has been serving not only as an effective medium not only to inform and educate people but also to promote their folk culture.

4. Radio breaks the literacy barrier: Radio is easily understandable to the literates and the non-literates. Unless you are literate, you cannot read a newspaper or read captions or text on television. But for listening to radio, you need not be literate. The illiterate person in a remote village can be a regular radio listener.

5. Radio is portable: Radio can be accessed on the move. The listener need not sit in one position to listen to the radio. You can listen to it while doing your work or while driving. It can accompany you and entertain you anywhere. Hence, it is a convenient medium.

6. Radio is a low cost medium: Radio is an inexpensive medium. The cost of production is low and a small radio can be bought for as low a price assay, fifty rupees. Before the advent of television, radio was the chief means of communication for people of all classes. However, not everyone owned the radio sets. Many people listened to one radio at the same time. Then transistor revolution spread the ownership of radio sets in a rapid way.

7. Radio does not need electric power supply: You can listen to radio using dry battery cells even if you do not have electric power supply or a generator. While people in cities spend their evenings watching television, in rural areas, where there is no electricity or erratic supply of electricity, people still prefer the radio for their entertainment.

8. Radio is a medium of immediacy: Radio can deliver messages instantly. It can be the first to report the happenings while TV crew would take some time to reach the spot.

As things happen in a studio or outside, messages can be sent or broadcast live. These messages can be picked up by anyone who has a radio set or receiver which is tuned into a radio station. Irrespective of our location, we can listen to radio in the language of our choice.

Limitations of Radio

Radio as a medium has its own limitations as well. Some of them are as follows:

- Radio entirely depends on the sense of hearing. It has no visual images. Unlike television, broadcast is not reinforced by the powerful medium of sight. For instance, it is difficult to convey the intricacies of works of art such as paintings, sculptures or handicrafts merely bywords. Also, in case of a a major disaster -say an earthquake or a war, it is easy to portray the extent of damage or the hardships faced by the people clearly on television. On radio, the listener has to use his imagination to pictures the situation in his mind. And there can be gaps between illusion and reality. Television, being a visual medium, has the advantage of communicating a message through facial expressions or body language. On radio both, the broadcaster and the listener, have to constantly keep in mind that what is being conveyed will have to be heard, understood and remembered instantly.

- Radio is an ephemeral medium, unless one has access to a recording or a repeat broadcast, the message can be lost forever. This puts pressure on the broadcaster to convey his message emphatically in one chance and demands a great deal of concentration on the listener's part as well. For example, while reading a newspaper, if you do not understand the meaning of certain words, you can refer to a dictionary to find out the meaning. You can refer to the article again and again. This is not possible for radio. While listening to a news bulletin on radio, if you refer to a dictionary or ask someone else for the meaning, you will miss the rest of the news. You have to understand what is being said on radio as you listen. What is said on radio does not exist any longer; unless you record it.

- Messages on radio can be easily forgotten. Sometimes, a visual on television or an article in a newspaper may have a lasting impact on a person.

- There is a lot of dependence on presentation in radio. If a presenter is boring, the listeners may lose interest in the programme. So not just the message but how it is presented too matters a lot in radio.

- Radio has little value for the hearing-challenged just as television is of little use to the visually-challenged.

Fax

In telecommunications, Fax (facsimile) is also called telefax, in which the transmission and reproduction of documents by wire or radio wave. Common fax machines are de-signed to scan printed textual and graphic material and then transmit the information through the telephone network to similar machines, where facsimiles are reproduced close to the form of the original documents. Fax machines, because of their low cost and their reliability, speed, and simplicity of operation, revolutionized business and personal correspondence. They virtually replaced telegraphic services, and they also present an alternative to government-run postal services and private couriers.

Fax machines send and receive information using a telephone line.

Standard Fax Transmission

Most office and home fax machines conform to the Group 3 standard, which was adopt-ed in 1980 in order to ensure the compatibility of digital machines operating through public telephone systems worldwide. As a standard letter-size sheet is fed through a machine, it is scanned repeatedly across its width by a charge-coupled device (CCD), a solid-state scanner that has 1,728 photo-sensors in a single row. Each photo-sensor in turn generates a low or high variation in voltage, depending on whether the scanned spot is black or white. Since there normally are 4 scan lines per mm (100 scan lines per inch), the scanning of a single sheet can generate almost two million variations in voltage.

The high/low variations are converted to a stream of binary digits, or bits, and the bit stream is subjected to a source encoder, which reduces or "compresses" the number of bits required to represent long runs of white or black spots. The encoded bit stream can then be modulated onto an analog carrier wave by a voice-band modem and transmit-ted through the telephone network. With source encoding, the number of bits required to represent a typewritten sheet can be reduced from two million to less than 400,000.

As a result, at standard fax modem speeds (up to 56,000 bits per second, though usually less) a single page can be transmitted in as little as 15 seconds.

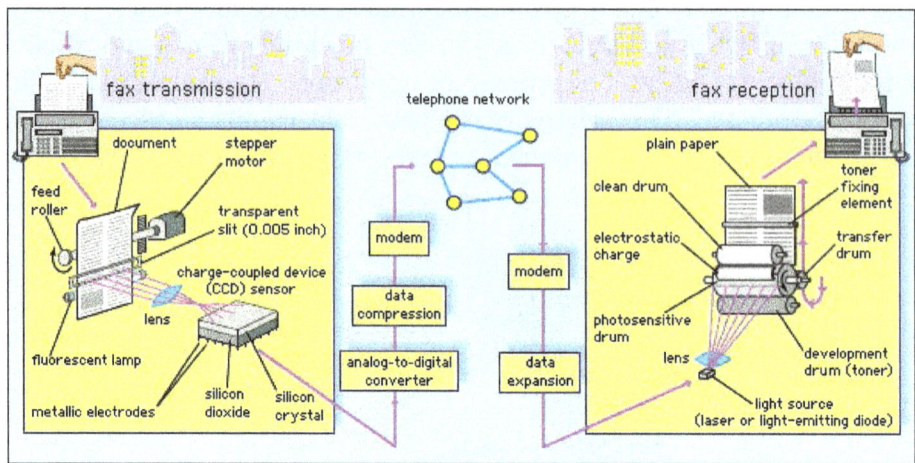

Digital fax transmission and reception, using a scanner and printer connected by modem to the public switched telephone network.

Communication between a transmitting and a receiving fax machine opens with the dialing of the telephone number of the receiving machine. This begins a process known as the "handshake," in which the two machines exchange signals that establish compatible features such as modem speed, source code, and printing resolution. The page information is then transmitted, followed by a signal that indicates no more pages are to be sent. The called machine signals receipt of the message, and the calling machine signals to disconnect the line.

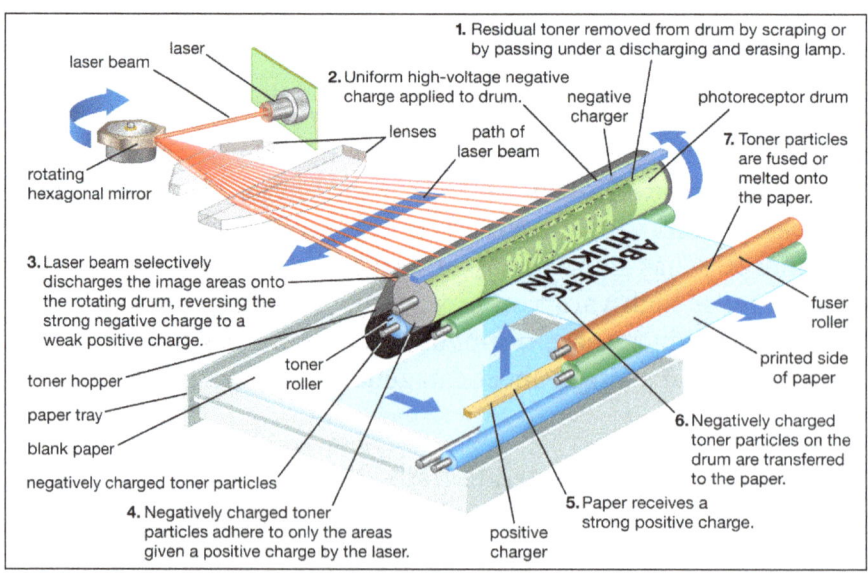

Laser printer.

At the receiving machine, the signal is demodulated, decoded, and stored for timed release to the printer. In older fax machines the document was reproduced on special thermally sensitive paper, using a print head that had a row of fine wires corresponding to the photo-sensors in the scanning strip. In modern machines it is reproduced on plain paper by a xerographic process, in which a minutely focused beam of light from a semiconductor laser or a light-emitting diode, modulated by the incoming data stream, is swept across a rotating, electrostatically charged drum. The drum picks up toner powder in charged spots corresponding to black spots on the original document and transfers the toner to the paper.

Group 3 facsimile transmission can be conducted through all telecommunications media, whether they be copper wire, optical fibre, microwave radio, or cellular radio. In addition, personal computers (PCs) with the proper hardware and software can send files directly to fax machines without printing and scanning. Conversely, documents from a remote fax machine may be received by a computer for storage in its memory and eventual reproduction on a desktop printer. Internet fax servers have been developed that can send or receive facsimile documents and transmit them by e-mail between PCs.

The concepts of facsimile transmission were developed in the 19th century using contemporary telegraph technology. Widespread employment of the method, however, did not take place until the 1980s, when inexpensive means of adapting digitized information to telephone circuits became common.

Early Telegraph Facsimile

Facsimile transmission over wires traces its origins to Alexander Bain, a Scottish mechanic. In 1843, less than seven years after the invention of the telegraph by American Samuel F.B. Morse, Bain received a British patent for "improvements in producing and regulating electric currents and improvements in timepieces and in electric printing and signal telegraphs." Bain's fax transmitter was designed to scan a two-dimensional surface (Bain proposed metal type as the surface) by means of a stylus mounted on a pendulum. The invention was never demonstrated.

Frederick Bakewell, an English physicist, was the first to actually demonstrate facsimile transmission. The demonstration took place in London at the Great Exhibition of 1851. Bakewell's system differed somewhat from Bain's in that images were transmitted and received on cylinders—a method that was widely practiced through the 1960s. At the transmitter the image to be scanned was written with varnish or some other non-conducting material on tinfoil, wrapped around the transmitter cylinder, and then scanned by a conductive stylus that, like Bain's stylus, was mounted to a pendulum. The cylinder rotated at a uniform rate by means of a clock mechanism. At the receiver a similar pendulum-driven stylus marked chemically treated paper with an electric current as the receiving cylinder rotated.

The first commercial facsimile system was introduced between Lyon and Paris, France, in 1863 by Giovanni Caselli, an Italian inventor. The first successful use of optical scanning and transmission of photographs was demonstrated by Arthur Korn of Germany in 1902. Korn's transmitter employed a selenium photocell to sense an image wrapped on a transparent glass cylinder; at the receiver the transmitted image was recorded on photographic film. By 1906 Korn's equipment was put into regular service for transmission of newspaper photographs between Munich and Berlin via telegraph circuits.

Analog Telephone Facsimile

Further deployment of fax transmission had to await the development of improved long-distance telephone service. Between 1920 and 1923 the American Telephone & Telegraph Company (AT&T) worked on telephone facsimile technology, and in 1924 the telephotography machine was used to send pictures from political conventions in Cleveland, Ohio, and Chicago to New York City for publication in newspapers. The telephotography machine employed transparent cylindrical drums, which were driven by motors that were synchronized between transmitter and receiver. At the transmitter a positive transparent print was placed on the drum and was scanned by a vacuum-tube photoelectric cell. The output of the photocell modulated a 1,800-hertz carrier signal, which was subsequently sent over the telephone line. At the receiver an unexposed negative was progressively illuminated by a narrowly focused light beam, the intensity of which corresponded to the output of the photoelectric cell in the transmitter. The AT&T fax system was capable of transmitting a 12.7-by-17.8-cm (5-by-7-inch) photograph in seven minutes with a resolution of 4 lines per mm (100 lines per inch).

Further advancements in fax technology occurred during the 1930s and '40s. In 1948 Western Union introduced its desk-fax service, which was based on a small office machine. Some 50,000 desk-fax units were built until the service was discontinued in the 1960s.

Over the years, different manufacturers adopted operability standards that allowed their machines to communicate with one another, but there was no worldwide standard that enabled American machines, for example, to connect to European fax machines. In 1974 the International Telegraph and Telephone Consultative Committee (CCITT) issued its first worldwide fax standard, known as Group 1 fax. Group 1 fax machines were capable of transmitting a one-page document in about six minutes with a resolution of 4 lines per mm using an analog signal format. This standard was followed in 1976 by a CCITT Group 2 fax standard, which permitted transmission of a one-page document in about three minutes using an improved modulation scheme.

Digital Facsimile

Although the Group 2 fax machines proved to be successful in business applications where electronic transmission of documents containing non-textual information such as drawings, diagrams, and signatures was required, the slow transmission rate and the cost of the terminals ultimately limited the growth of fax services. In response, the CCITT developed standards for a new class of fax machine, now known as Group 3, which would use digital transmission of images through modems. With the encoding of a scanned image into binary digits, or bits, various image-compression methods (also known as source encoding or redundancy reduction) could be employed to reduce the number of bits required to represent the original image. By coupling a good source code with a high-speed modem, a Group 3 fax machine could reduce the time required to transmit a single page to less than one minute—a threefold improvement in transmission time over the older Group 2 fax machines. The Group 3 standard was adopted by the CCITT in 1980.

Originally, Group 3 fax was intended for transmission at data rates between 2,400 and 9,600 bits per second. With advances in voice-band modem technology, data transmission rates of 28,800 bits per second and above became common. Between 1981 and 1984 the CCITT sponsored the development of a high-speed fax service that was adopted as the Group 4 standard in 1984. Group 4 fax was intended to supplant Group 3 fax by permitting error-free transmission of documents over digital networks, such as the integrated services digital network (ISDN), at speeds up to 64,000 bits per second. At such rates, transmission time for a single page could be reduced to less than 10 seconds. Group 4 fax has been deployed in areas of the world where ISDN lines are readily available (e.g., Japan and France). However, since other areas (e.g., the United States) do not have many ISDN lines installed in the local telephone loop, Group 4 fax machines must also support Group 3 fax for transmission over analog lines.

Email

Email (electronic mail) is a way to send and receive messages across the Internet. It's similar to traditional mail, but it also has some key differences.

Email Advantages

- Productivity tools: Email is usually packaged with a calendar, address book, instant messaging, and more for convenience and productivity.

- Access to web services: If you want to sign up for an account like Facebook or order products from services like Amazon, you will need an email address so you can be safely identified and contacted.

- Easy mail management: Email service providers have tools that allow you to file, label, prioritize, find, group, and filter your emails for easy management. You can even easily control spam, or junk email.

- Privacy: Your email is delivered to your own personal and private account with a password required to access and view emails.

- Communication with multiple people: You can send an email to multiple people at once, giving you the option to include as few as or as many people as you want in a conversation.

- Accessible anywhere at any time: You don't have to be at home to get your mail. You can access it from any computer or mobile device that has an Internet connection.

Understanding Email Addresses

To receive emails, you will need an email account and an email address. Also, if you want to send emails to other people, you will need to obtain their email addresses. It's important to learn how to write email addresses correctly because if you do not enter them exactly right, your emails will not be delivered or might be delivered to the wrong person.

Email addresses are always written in a standard format that includes a user name, the @ (at) symbol, and the email provider's domain. The user name is the name you choose to identify yourself.

Merced Flores · merced.flores73 @gmail.com>
to me
Hi Julia,

The email provider is the website that hosts your email account.

Merced Flores <merced.flores73@ gmail.com·
to me
Hi Julia,

Some businesses and organizations use email addresses with their own website domain.

from: Facebook <security@ facebookmail.com·

About Email Providers

People usually received an email account from the same companies that provided their Internet access. For example, if AOL provided your Internet connection, you'd have an AOL email address. While this is still true for some people, today it's increasingly common to use a free web-based email service, also known as webmail. Anyone can use these services, no matter who provides their Internet access.

Webmail Providers

Today, the top three webmail providers are Yahoo!, Microsoft's Outlook.com (previously

Hotmail), and Google's Gmail. These providers are popular because they allow you to access your email account from anywhere with an Internet connection. You can also access webmail on your mobile device.

Other Email Providers

Many people also have an email address hosted by their company, school, or organization. These email addresses are usually for professional purposes. For example, the people who work for this website have email addresses that end with @gcflearnfree.org. If you are part of an organization that hosts your email, they'll show you how to access it.

Many hosted web domains end with a suffix other than .com. Depending on the organization, your provider's domain might end with a suffix like .gov (for government websites), .edu (for schools), .mil (for military branches), or .org (for non-profit organizations).

Email Productivity Features

In addition to email access, webmail providers offer various tools and features. These features are part of a productivity suite—a set of applications that help you work, communicate, and stay organized. The tools offered will vary by provider, but all major webmail services offer the following features: In addition, each provider offers some unique features.

For instance, when you sign up for Gmail you gain access to a full range of Google services, including Google Drive, Google Docs, and more. Outlook, on the other hand, offers connectivity with OneDrive and Microsoft Office Web Apps.

Getting started with Email

You should now have a good understanding of what email is all about. Over the next few lessons, we will continue to cover essential email basics, etiquette, and safety tips.

Setting up your Own Email Account

If you want to sign up for your own email account, we suggest choosing from one of the three major webmail providers.

Practice using an Email Program

Keep in mind that this tutorial will not show you how to use a specific email account. It's a useful course for learning the basics, even if you ultimately end up choosing an email provider other than Gmail, such as Yahoo! or Outlook.com. There, you will learn how to:

- Sign up for an email account.

- Navigate and get to know the email interface.

- Compose, manage, and respond to email.

- Set up email on a mobile device.

Telephony

Telephony is a term denoting the technology that allows people to have long distance voice communication. The term's scope has been broadened with the advent of new communication technologies. In its broadest sense, the terms encompass phone communication, Internet calling, mobile communication, faxing, voicemail and even video conferencing. It is finally difficult to draw a clear line delimiting what is telephony and what isn't.

The initial idea that telephony returns to is the POTS (plain old telephone service), technically called the PSTN (public switched telephone network). This system is being fiercely challenged by and to a great extent yielding to Voice over IP (VoIP) technology, which is also commonly referred to as IP Telephony and Internet Telephony.

Voice Over IP (VoIP) and Internet Telephony

These two terms are used interchangeably in most cases, but technically speaking, they are not quite the same thing. The three terms that personate one another are Voice over IP, IP Telephony, and Internet Telephony. They all refer to the channeling of voice calls and voice data through IP networks, namely LANs and the Internet. This way, existing facilities and resources that are already used for data transmission are harnessed, thereby eliminating the cost of expensive line dedication as is the case with the PSTN. The main advantage that VoIP brings to users is considerable cost-cutting. Calls are also often free.

This along with the numerous advantages that VoIP brings has caused the latter to become a major technological element that has gained worldwide popularity and claimed the lion's share of the telephony market. The term Computer Telephony has emerged with the advent of softphones, which are applications installed on a computer, mimicking a phone, using VoIP services on the Internet. Computer telephony has become very popular because most people use it for free.

Mobile Telephony

Who doesn't carry telephony in their pocket nowadays? Mobile phones and handsets normally use mobile networks using the GSM (cellular) technology to allow you to make calls on the move. GSM calling is rather expensive, but VoIP has also invaded mobile phones, smartphones, pocket PCs and other handsets, allowing mobile users to make very cheap and sometimes free local and international calls. With mobile VoIP, Wi-Fi, 3G, 4G, and newer technologies allow users to make completely free calls, even to overseas contacts.

Telephony Equipment and Requirements

What is required for telephony ranges between very simple hardware to complex equipment? Let us stay on the client-side (your side as a customer) so as to avoid the complexities of PBXs and servers and exchanges.

For PSTN, you only need a phone set and a wall jack. With VoIP, the main requirement is a connection to either an IP network (e.g. an Ethernet or Wi-Fi, connection to a LAN), a broadband Internet connection and, in the case of mobile telephony, a wireless network connection like Wi-Fi, 3G, and in some cases GSM. The equipment can then be as simple as a headset (for computer telephony). For those that want the convenience of the home phone without the computer, they need an ATA (also called a phone adapter) and a simple traditional phone. An IP phone is a special phone that includes the functionality of an ATA and many other features and therefore can work without depending on other hardware.

Not Only Voice

Since many media mix up on one channel, faxing and video conferencing also fall under the telephony banner. Faxing traditionally uses the phone line and phone numbers to transmit facsimile (shortened to fax) messages. IP Faxing uses IP networks and the Internet to send and receive fax messages. This gives many advantages but still faces certain challenges. Video conferencing works the same way as voice over IP with added real-time video.

Digital Telephony

Digital telephony is used in the provision of digital telephone services and systems. One of the primary challenges when transmitting analog signals is that all sorts of things can interfere with those signals, causing low volume, static, and all manner of other undesired effects.

This is the principle of all digital audio (including telephony): Sample the characteristics of the source waveform, store the measured information, and send that data to the far end. Then, at the far end, use the transmitted information to generate a completely new audio signal that has the same characteristics as the original. The reproduction is so good that the human ear can't tell the difference.

The principal advantage of digital audio is that the sampled data can be mathematically checked for errors all along the route to its destination, ensuring that a perfect duplicate of the original arrives at the far end. Distance no longer affects quality, and interference can be detected and eliminated.

Pulse-code Modulation

There are several ways to digitally encode audio, but the most common method (and the one used in telephony systems) is known as Pulse-Code Modulation (PCM). To illustrate how this works, let's go through a few examples.

Digitally Encoding an Analog Waveform

The principle of PCM is that the amplitude of the analog waveform is sampled at specific intervals so that it can later be re-created. The amount of detail that is captured is dependent both on the bit resolution of each sample and on how frequently the samples are taken. A higher bit resolution and a higher sampling rate will provide greater accuracy, but more bandwidth will be required to transmit this more detailed information.

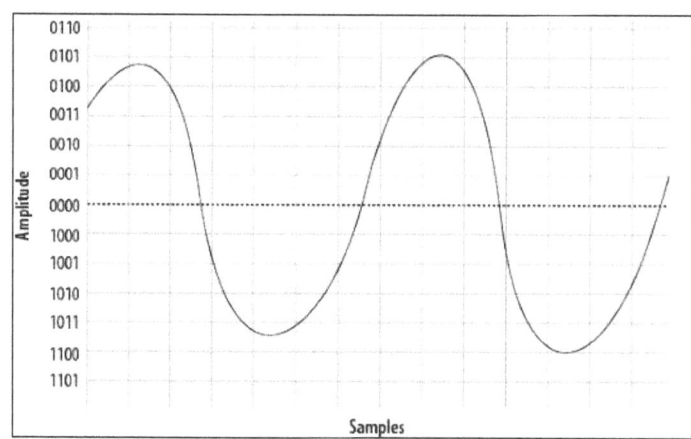

A simple sinusoidal (sine) wave.

To digitally encode the wave, it must be sampled on a regular basis, and the amplitude of the wave at each moment in time must be measured. The process of slicing up a waveform into moments in time and measuring the energy at each moment is called quantization, or sampling.

The samples will need to be taken frequently enough and will need to capture enough information to ensure that the far end can re-create a sufficiently similar waveform. To achieve a more accurate sample, more bits will be required. To explain this concept, we will start with a very low resolution, using 4 bits to represent our amplitude. This will make it easier to visualize both the quantization process itself and the effect that resolution has on quality.

Pros and Cons of Digital Telephony

Users of modern digital telephony are supported by any type of technology, whether Digital Enhanced Cordless Telecommunications (DECT), Wireless Local Area Network (WLAN), Internet Protocol (IP), Time Division Multiplexing (TDM) or Session Initiation Protocol (SIP), functions of a regular phone, and more importantly – the opportunity to transmit voice, video or data. In this case, because if you call with IP-phones is not activated international (national carrier), the cost of such calls is significantly lower than with traditional phone call.

Another important advantage is that – safety. Although most consumers are not IP Telephony but it is relatively easy to install plays such calls, or even change their content, but all are the introduction of VoIP protocol can significantly complicate the interference in the system. In addition, although absolutely secure methods of communication and does not exist, it is in IP telephony which can produce complex multi-level encryption, providing complete anonymity of all VoIP traffic with I2P Network (router software).

When using IP Telephony, coding speech information leads to distortion and deterioration of the quality of communication and speech. Experts believe that this happens in case of packet loss at the time of transmission networks or due to slow Internet speed, as well as if it involved potentially exceeding the time required for delivery of a package that carries voice data. All this requires careful work to optimize the delay in the network and algorithms for recovering lost.

Terrestrial Television

Terrestrial television is the traditional method of television broadcast signal delivery. Terrestrial television broadcasting dates back to the very beginnings of television as a medium itself with the first long distance public television broadcast from Washington, DC on April 7, 1927. There was virtually no other method of television delivery until the advent of cable television, or community antenna television (CATV) in the 1950s. The first non-terrestrial method of delivering television signals began with the use of communications satellites during the 1960s and 1970s. In the Indian context, cable television networks made their appearance in early 1990s and the first Direct to Home (DTH) Satellite television service was launched in October 2003.

Apart from cable television networks and DTH service, analog terrestrial television is now also subject to competition from distribution of video and film content over the

Internet and telecommunication networks. The technology of digital terrestrial television has evolved as a response to these challenges.

Analogue Terrestrial Television

Europe

Following the ST61 conference, UHF frequencies were first used in the UK in 1964 with the introduction of BBC2. In UK, VHF channels were kept on the old 405-line system, while UHF was used solely for 625-line broadcasts (which later used PAL colour). Televisions broadcasting in the 405-line system continued after the introduction of four analogue programmes in the UHF bands until the last 405-line transmitters were switched off on January 6, 1985. VHF Band III was used in other countries around Europe for PAL broadcasts until planned phase-out and switch over to digital television.

The success of analogue terrestrial television across Europe varied from country to country. Although each country had rights to a certain number of frequencies by virtue of the ST61 plan, not all of them were brought into service.

Rooftop television antennas like these are required to receive terrestrial television in fringe reception areas far from the television station.

Americas

In 1941, the first NTSC standard was introduced by the National Television System Committee. This standard defined a transmission scheme for a black and white picture with 525 lines of vertical resolution at 60 fields per second. In the early 1950s, this standard was superseded by a backwards-compatible standard for color television. The NTSC standard was exclusively being used in the Americas as well as Japan until the introduction of digital terrestrial television (DTT). While Mexico have ended all its analogue television broadcasts and the US and Canada have shut down nearly all of their analogue TV stations, the NTSC standard continues to be used in the rest of Latin American countries while testing their DTT platform.

In the late 1990s and early 2000s, the Advanced Television Systems Committee developed the ATSC standard for digital high definition terrestrial transmission. This standard was eventually adopted by many American countries, including the United States, Canada, Dominican Republic, Mexico, El Salvador, Guatemala and Honduras; however, the three latter countries ditched it in favour of ISDB-Tb.

The Pan-American terrestrial television operates on analog channels 2 through 6 (VHF-low band, 54 to 88 MHz, known as band I in Europe), 7 through 13 (VHF-high band, 174 to 216 MHz, known as band III elsewhere), and 14 through 51 (UHF television band, 470 to 698 MHz, elsewhere bands IV and V). Unlike with analog transmission, ATSC channel numbers do not correspond to radio frequencies. Instead, a virtual channel is defined as part of the ATSC stream metadata so that a station can transmit on any frequency but still show the same channel number. Additionally, free-to-air television repeaters and signal boosters can be used to rebroadcast a terrestrial television signal using an otherwise unused channel to cover areas with marginal reception.

Analog television channels 2 through 6, 7 through 13, and 14 through 51 are only used for LPTV translator stations in the U.S. Channels 52 through 69 are still used by some existing stations, but these channels must be vacated if telecommunications companies notify the stations to vacate that signal spectrum. By convention, broadcast television signals are transmitted with horizontal polarization.

Asia

Terrestrial television broadcast in Asia started as early as 1939 in Japan through a series of experiments done by NHK Broadcasting Institute of Technology. However, these experiments were interrupted by the beginning of the World War II in the Pacific. On February 1, 1953, NHK (Japan Broadcasting Corporation) began broadcasting. On August 28, 1953, Nippon TV (Nippon Television Network Corporation), the first commercial television broadcaster in Asia was launched. Meanwhile, in the Philippines, Alto Broadcasting System (now ABS-CBN Corporation), the first commercial television broadcaster in Southeast Asia, launched its first commercial terrestrial television station DZAQ-TV on October 23, 1953, with the help of Radio Corporation of America (RCA).

Digital Terrestrial Television

Digital Terrestrial Television (DTT) is the most common type of TV service across the world. In the UK it is known as Free view and it replaced the old analogue TV service which consisted of five channels. With Free view you can get up 70 free-to-air standard channels, 15 HD channels and around 30 radio services. TV transmitters send digital TV signals through the air which are received by an aerial, often on a rooftop or in the loft of your home, which is connected to the TV.

Most cities and most towns across the UK are served by large 'main' transmitters. These are large structures which send TV signals to many millions of households. Areas that can't be covered by a main transmitter (for example where a hill blocks the signal) are normally served by smaller 'relay' transmitters. Main transmitters broadcast a larger number of channels compared to relay transmitters. All transmitters carry the most popular channels.

Instant Messaging

Instant messaging, often shortened to IM or IM'ing, is the exchange of near real-time messages through a stand-alone application or embedded software. Unlike chat rooms with many users engaging in multiple and overlapping conversations, IM sessions usually take place between two users in a private, back-and-forth style of communication.

One of the core features of many instant messenger clients is the ability to see whether a friend or co-worker is online and connected through the selected service a capability known as presence. As the technology has evolved, many IM clients have added support for exchanging more than just text-based messages, allowing actions like file transfers and image sharing within the IM session.

Instant messaging differs from email in the immediacy of the message exchange. IM also tends to be session-based, having a start and an end. Because IM is intended to mimic in-person conversations, individual messages are often brief. Email, on the other hand, usually reflects a longer-form, letter-writing style.

How Instant Messaging Works

Generally, IM users must know each other's username or screen name to initiate an IM session or to add them to their contact list or buddy list. Once the intended recipient has been identified and selected, the sender opens an IM window to begin the session.

For IM'ing to work as intended, both users must be online at the same time, although nearly all instant messaging platforms now allow asynchronous interactions between online and offline users. If offline messaging is not supported, attempting to IM an unavailable user will result in a notification that the transmission cannot be completed. In addition, the intended recipient must be willing to accept instant messages, as it is possible to configure the IM client to reject certain users.

When an IM is received, it alerts the recipient with a window containing the incoming message. Or, depending on the user's settings, a window could indicate an IM has arrived along with a prompt to accept or reject it. Many IM clients also notify the user audibly with a distinctive sound, such as a chime or chirp. The user can also be notified visually by flashing the IM window or its taskbar icon when a message has arrived.

While IM clients were frequently based on proprietary protocols in the past requiring both users to use the same software in order to communicate the adoption of open standards has become more common. This has enabled the rise of multi-platform instant messengers, such as Pidgin and Trillian.

Another important shift in IM has been how it's accessed and delivered. Long deployed as a desktop client that had to be downloaded and installed, instant messaging is now more often found as a feature within another web- or cloud-based service such as Facebook, Gmail and Skype or as a mobile app, such as WhatsApp Messenger.

Features of Instant Messaging

The exchange of text has long been the chief function of instant messaging, but it is now one feature of many. The ability to insert images and emojis into messages is now standard in many clients, as are file transfers. Facebook Messenger even enables users to send money via IM. Numerous clients now support the escalation from IM to other modes of communication, such as group chat, voice calls or video conferencing.

Presence enables users to see the availability of their contacts -- not only whether they are online or offline, but also whether they have indicated their status is free or busy. Some clients also enable users to set an "away message" providing more detail about their limited or lack of availability. Within an active session between two users, most clients can also indicate to one user in real time when the other user is typing.

Advantages of Instant Messaging

- Allows you to chat in 'real time' to other people who also have an IM client.

- IM allows you to get on with other things and yet be in touch real time with connected friends and colleagues.

- Useful for customer support contact instead of having to phone a support line.

Disadvantages of Instant Messaging

- As it is immediate, you have no time to reflect on the message you are sending, unlike an email where you can review the draft before sending.

- In order to provide a free service, the IM providers send adverts and popup windows to each person. If you want to avoid this, you need to pay for a 'premium' service.

- Unless you set up your IM client carefully, anyone can send you a message - not always a good thing.

Videoconferencing

A video conference is an online meeting (or a meeting over distance) that takes place between two parties, where each participant can see an image of the other, and where both parties are able to speak and listen to the other participants in real time. The components necessary to make this happen include:

- A microphone, webcam and speakers.

- A display.

- A software program that captures the voice stream from the microphone, encodes it, transmits to the other participant, and simultaneously decodes the digital voice stream being received from the remote participant in the video conference (most commonly referred to as a "Codec").

- A software program that bridges both parties together across a digital connection, managing the exchange of voice and video between participants. At either end of the connection, the video and voice traffic is combined and delivered to each participant in the form of a real-time video image and audio stream.

- An optional management tool for the scheduling of video conferencing sessions.

At a slightly more advanced level, it is also possible to provide the ability to share content from a device during a video call. The quality and type of content that can be shared depends on the rate of data exchange during the call.

Terminology used by video conferencing users to describe the process of dialing into and participating in a virtual meeting is known as "joining a bridge." Different virtual meeting rooms are assigned unique "bridge numbers," and users join a video call by "dialing a bridge number."

Point-to-point Video Conferencing

Video-enabled meetings happen in two distinct ways: either point-to-point or with multipoint. In point-to-point, the simplest scenario is where one person or group is connected to another. The physical components (i.e. microphone and camera) that enable the meeting to take place are often integrated in to desktop computing solutions like a laptop or tablet, or can be combined into dedicated, room-based hardware solutions.

Where desktop solutions tend to be used by individuals, room-based solutions utilize dedicated video conferencing technology where groups of people can be seen, heard and can naturally participate in the meeting.

An example of point-to-point video conferencing.

Multi-Point Video Conferencing

In multi-point video calls, three or more locations are connected together, where all participants can see and hear each other, as well as see any content being shared during the meeting. In this scenario, digital information streams of voice, video and content are processed by a central, independent software program. Combining the individual participant's video and voice traffic, the program re-sends a collective data stream back to meeting participants in the form of real-time audio and video imagery.

Individuals can participate in a meeting in an "audio only" mode, or combine audio with video images of the meeting on screen. Depending upon the technical capability of the video conferencing system being used, images seen by participants are either classified as "active speaker" or "continuous presence."

A use-case scenario of multi-point video conferencing.

In "active speaker" mode, the screen only provides an image of the person that is speaking at any point in time. In more advanced solutions with "continuous presence" mode, the bridge divides the image on the screen into a number of different areas. The person speaking at any point in time is presented in a large central area, and other meeting participants are shown displayed around the central image.

The "continuous presence" mode thus allows meeting participants to view and interact with all meeting participants in a 'virtual meeting room.' The software program which creates the "virtual meeting room" and the digital processing hardware, on which it resides, is often called a video bridge, or "bridge", for short. Another term for a bridge which is often used is a video conferencing "multi-point control unit" or "MCU."

Whereas point-to-point video conferencing is relatively simple, the creation and management of multi-point video conferences can be complex. An MCU must be able to create, control and facilitate multiple simultaneous live video conferencing meetings. A further complexity is added when different locations may connect to the meeting over digital or analogue streams at different speeds, with different data transport and signaling protocols employed to facilitate the communication.

To link these users into a common, virtual meeting, the MCU must therefore be able to understand and translate between several different protocols (i.e. H.264 for communication over IP, and H.263 for ISDN). The MCU will also allow those joining the video bridge to do so at the highest speed and the best possible quality that their individual system can support. Although there are two separate processes taking place here, this is often jointly referred to as "transcoding."

It is important to note that not all bridges provide such transcoding capability, and failure to do this can seriously impact the quality and experience of video calls. When transcoding is not provided and users dial into a bridge over a range of different connection speeds, it is possible that the bridge may only be able to support the video meeting by establishing the connections at the lowest common denominator. To illustrate the negative effect of this, consider a meeting that takes place with most users joining the bridge from the high-speed corporate network, but where one or two individuals dial into the meeting from home on low-bandwidth DSL or ISDN. In this case the experience of the many corporate users is downgraded to the lowest common denominator of the home-users, potentially making the video call ineffective. Where effective transcoding is supported by the MCU, those on the corporate network will continue to enjoy HD video quality, while remote users receive quality commensurate with their connection speeds.

In summary, when an MCU is designed well, integrating easily with multiple vendors and allowing users to call in at the data rate and resolution they want or need to—the result is an easy, seamless experience for all users, allowing people to focus on the meeting, not the technology.

The Language of Video Conferencing

As video conferencing technology has evolved, two main protocols have emerged to provide the signaling control for the establishment, control and termination of video conferencing calls: SIP (Session Initiation Protocol) and H.323.

For the encoding and decoding of visual information, the industry is moving towards the industry standard known as H.264, which was developed to provide high-quality video at lower bandwidth over a wide range of networks and systems. An extension to the H.264 protocol is Scalable Video Coding (SVC), which is established to facilitate the enablement of video conferencing on a wider range of devices, such as tablets and mobile phones.

Bridging Architecture and Functionality

The combination of software and the hardware that creates the virtual meeting rooms is called a "video bridge." Virtual meeting rooms are identified by their "bridge numbers." With multiple calls taking place simultaneously, software analyses all the different data streams coming into the bridge processors, and assigns data streams accordingly. At the simplest level, the processing workload for bridges is dependent upon four factors:

- The number of locations that dial into each bridge.

- The number of conferencing calls that each bridge must handle simultaneously.

- The amount of data that is being received on each digital stream: higher resolutions of images and sound (i.e. High Definition) generate more data that needs to be processed.

- The degree of transcoding that the bridge must perform while handling calls being received at different connection speeds and utilizing different protocols.

As the workload increases, each bridge must process more data. Performance can therefore be improved by increasing the number of Digital Signaling Processors (DSPs) utilized to decode and encode the digital streams entering and leaving MCUs. If the bridging function becomes overloaded, video and voice information may be lost, causing latency to be introduced into calls, both of which can degrade the video meeting experience.

Extra processing resource can be provided for the bridging function by either utilizing a more powerful bridge (with a greater number of DSPs) or through virtual software approach, where the software that controls the signaling function can operate independently of the physical hardware.

A conference call with an assigned conference number does not have to take place, or be processed by a dedicated piece of hardware. The call can be "virtualized", and assigned to whatever physical bridge has the correct resource or capacity to handle the call. A virtualization manager oversees which physical bridge has the capacity, and assigns incoming calls accordingly.

In extreme, but rare circumstances, the virtualization manager may assign resources for a call across several different physical bridges that work in tandem together. Known as "auto-cascading", the resources within the physical bridge can be instructed by the software to operate in a "parent-child" arrangement, with one bridge "owning" the conference call, and the others sharing the workload.

In the continuous presence mode of presentation, the bridge will automatically provide the screen templates in which the viewers will see the other meeting participants. The bridge can also provide some administrative functionality for the call, such as assigning passwords to enter each meeting, and providing Interactive Voice Response (IVR) functionality, where call participants can be greeted and instructed by customized voice greetings.

Although most participants will actively dial into a video conferencing meeting, the bridge can be programmed to automatically dial out to participating locations and automatically connect them in to a meeting. For example, the bridge could automatically wake up the cameras in remote meeting rooms, and link those meeting rooms into a prescheduled call. Participants of such a meeting would simply have to walk into the video room at the correct time, and join the meeting.

Video Call Management and Protocol Conversion

In order to build an architecture that scales, the software platform must be able to provide call signaling functionality, and dynamically manage the set-up and maintenance of a large number of video calls. The software architecture has to be capable of reconfiguring itself and its resources in real-time, so that these resources are used to their best ability. In addition, the software architecture has to understand the bandwidth requirements of each call being placed, the policy that is associated with each call (the prioritization and importance of a call), and where the participants of a call are geographically located. By understanding this, the software platform can utilize local resources instead of redirecting data streams and call signaling to resources that are far away, an approach which would eat up large amounts of bandwidth on WAN links that are very costly.

The software platform should also be able to instantly detect any failure of hardware resources or loss of communication across infrastructure links, so that it can re-direct traffic and re-establish calls utilizing alternative resources, without overly impacting video calls or their quality.

When systems on different customer premises try to join the same video call using devices which run different protocols (i.e. H.323, RTV or SIP), the video conferencing platform must first perform protocol conversion to a common language so the infrastructure can understand and process information correctly. In other words, the software platform should provide intrinsic gateway functionality between devices that talk different languages.

Elements of a Telecommunication System

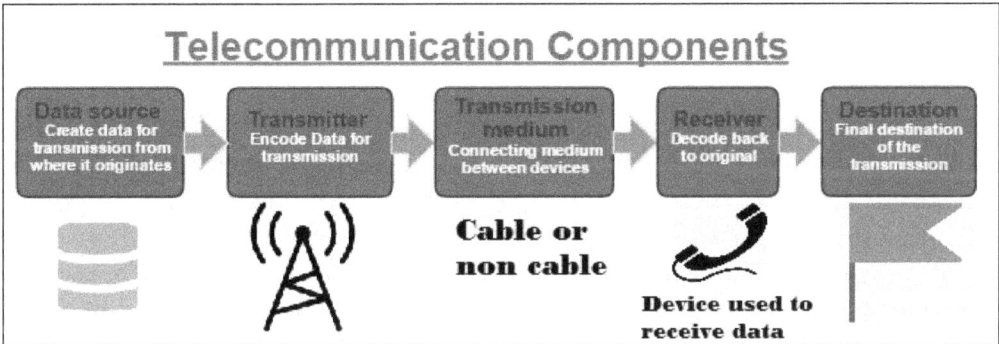

Transmitter

Transmitter (information source) that takes information and converts it to a signal for transmission. In electronics and telecommunications a transmitter or radio transmitter is an electronic device which, with the aid of an antenna, produces radio waves. In addition to their use in broadcasting, transmitters are necessary component parts of many electronic devices that communicate by radio, such as cell phones.

Transmission Medium

Transmission medium over which the signal is transmitted. For example, the transmission medium for sounds is usually air, but solids and liquids may also act as transmission media for sound. Many transmission media are used as communications channel. One of the most common physical Medias used in networking is copper wire. Copper wire to carry signals to long distances using relatively low amounts of power. Another example of a physical medium is optical fiber, which has emerged as the most commonly used transmission medium for long-distance communications. Optical fiber is a thin strand of glass that guides light along its length. The absence of a material medium in vacuum may also constitute a transmission medium for electromagnetic waves such as light and radio waves.

Receiver (Radio)

Receiver (information sink) that receives and converts the signal back into required information. In radio communications, a radio receiver is an electronic device that receives radio waves and converts the information carried by them to a usable form. It is used with an antenna. The information produced by the receiver may be in the form of sound (an audio signal), images (a video signal) or digital data.

Wired Communication

Wired communications make use of underground communications cables (less often, overhead lines), electronic signal amplifiers (repeaters) inserted into connecting cables at specified points, and terminal apparatus of various types, depending on the type of wired communications used.

Wireless

Wireless communication involves the transmission of information over a distance without help of wires, cables or any other forms of electrical conductors. Wireless operations permit services, such as long-range communications, that are impossible or impractical to implement with the use of wires. The term is commonly used in the telecommunications industry to refer to telecommunications systems (e.g. radio transmitters and receivers, remote controls etc.) which use some form of energy (e.g. radio waves, acoustic energy, etc.) to transfer information without the use of wires. Information is transferred in this manner over both short and long distances.

Communication Channel

For data to be communicated from one location to another, a physical pathway or medium must be used. These pathways are called communications media (channels) and can be either physical or wireless. The physical transmission use wire, cable, and other tangible materials; wireless transmission media send communications signals through the air or space. The physical transmission media are generally referred to as cable media (e.g., twisted pair wire, coaxial cable, and fiber optic cable). Wireless media include cellular radio, microwave transmission, satellite transmission, radio and infrared media.

Cable Media

Cable media (also called wire line media) use physical wires or cables to transmit data and information. Twisted-pair wire and coaxial cable are made of copper, and fiber-optic cable is made of glass. However, with the exception of fiber-optic cables, cables present several problems, notably the expense of installation and change, as well as a fairly limited capacity. Several cable media exist, and in many systems a mix of media (e.g., fiber coax) can be found.

Twisted-Pair Wire

Twisted-pair wire is the most prevalent form of communications wiring, because it is used for almost all business telephone wiring. Twisted-pair wire consists of strands of insulated copper wire twisted in pairs to reduce the effect of electrical noise.

Twisted-pair wire is relatively inexpensive, widely available, easy to work with, and can be made relatively unobtrusive by running it inside walls, floors, and ceilings. However, twisted-pair wire has some important disadvantages. It emits electromagnetic interference, is relatively slow for transmitting data, is subject to interference from other electrical sources, and can be easily "tapped" to gain unauthorized access to data.

Channel	Advantages	Disadvantages
Twisted-pair	• Inexpensive. • Widely available. • Easy to work with Unobtrusive.	• Slow (low bandwidth). • Subject to interference. • Easily tapped (low security).
Coaxial cable	• Higher bandwidth than twisted pair. • Less susceptible to electromagnetic interference.	• Relatively expensive and inflexible. • Easily tapped (low-to-medium security). • Somewhat difficult to work with. • Difficult to work with (difficult to splice).
Fiber-optic cable	• Very high bandwidth. • Relatively Inexpensive. • Difficult to tap (good security).	• Difficult to work with (difficult to splice).
Microwave	• High bandwidth. • Relatively inexpensive.	• Must have unobstructed line of sight. • Susceptible to environmental interference.
Satellite	• High bandwidth. • Large coverage area.	• Expensive. • Must have unobstructed line of sight. • Signals experience propagation delay. • Must use encryption for security.
Radio	• High bandwidth. • No wires needed Susceptible to snooping unless encrypted. • Signals pass through walls. • Inexpensive and easy to install.	• Create electrical interference problems.
Cellular Radio	• Low-to-medium bandwidth. • Signals pass through walls.	• Require construction of towers. • Susceptible to snooping unless encrypted.
Infrared	• Low-to-medium bandwidth.	• Must have unobstructed line of sight. • Used only for short distances.

Coaxial Cable

Coaxial cable consists of insulated copper wire surrounded by a solid or braided metallic shield and wrapped in a plastic cover. It is much less susceptible to electrical interference and can carry much more data than twisted-pair wire. For these reasons, it is commonly used to carry high-speed data traffic as well as television signals (i.e., in cable television). However, coaxial cable is 10 to 20 times more expensive, more difficult to work with, and relatively inflexible. Because of its inflexibility, it can increase the cost of installation or re-cabling when equipment must be moved. Data transmission over coaxial cable is divided into two basic types:

- Baseband: Transmission is analog, and each wire carries only one signal at a time.

- Broadband: Transmission is digital, and each wire can carry multiple signals simultaneously.

Because broadband media can transmit multiple signals simultaneously, it is faster and better for high-volume use. Therefore, it is the most popular Internet-access method. Broadband needs a network interface card (NIC), also called a LAN adapter, in order to run. An NIC is a card that is inserted into an expansion slot of computer or other device, enabling the device top connects to a network.

Fiber Optics

Fiber-optic cables (FOCs) are steadily replacing copper wire as a means of communications signal transmission. For example, Time Warner Telecom had 20,928 fiber route miles at the end of the first quarter of 2006, or capacity for its business customers to send 5,100 four-minute audio or video files at the same moment.

FOCs are used over long distances to connect local phone systems; and they provide the backbone for many network systems. Other FOC users are office buildings, industrial plants, cable TV services, university campuses, and electric utilities. Fiber is the ultimate medium for broadband (short for broad bandwidth). Bandwidth refers to the size of existing fiber-optic lines and their ability to carry all the data traffic companies want to send.

The fiber-optic system is similar to the copper wire system that it continues to replace. The key difference is that fiber optics use light pulses (light waves) to transmit information down fiber lines instead of using electronic pulses to transmit information down copper lines. The advantages of FOC over copper wire are:

- Speed: FOC networks operate at higher speeds, in the gigabits (Gbit). Industry forecasts indicate the 8-Gbit-per-second (Gbit/sec) fiber-optic channels will be widespread by 2011.

- Bandwidth: Larger carrying capacity. Using wavelength division multiplexing (WDM), the bandwidth carried by a single fiber is in the range of terabits per second. In comparison, the bandwidth for WiMax is in the one- megabit to two megabit range.

- Distance: Signals can be transmitted farther distances without needing to be strengthened (regenerated).

- Maintenance: FOC costs much less to maintain.

- Resistance: Greater resistance to electromagnetic noise such as radios, motors, or other nearby cables.

Telecommunication applications of fiber optics range from global networks to desktop computers. These involve the transmission of voice, data, or video over distances of less than a meter to hundreds of kilometers. Telecommunications carriers use optical fiber to carry plain old telephone service (POTS) across their nationwide networks. Local exchange carriers (LECs) use fiber to carry this same service between central office switches at local levels, and sometimes to neighborhoods or individual home. Fiber to the home (FTTH) is being deployed in select areas of the United States.

Multinational firms use FOC for secure, reliable data transfer between buildings to the desktop computers and to transfer data around the world. Cable television companies also use fiber for delivery of digital video and data services. The high bandwidth provided by fiber makes it the perfect choice for transmitting broadband signals, such as high-definition television (HDTV) telecasts. Intelligent transportation systems, such as smart highways with intelligent traffic lights, automated tollbooths, and changeable message signs, also use fiber-optic-based telemetry systems. The biomedical industry uses fiber-optic systems in modern telemedicine devices for transmission of digital diagnostic images. Other industries that use FOC extensively are space, military, automotive, and the industrial sector.

Wireless Media

Cable media (with the exception of fiber-optic cables) present several problems, notably the expense of installation and change, as well as a fairly limited capacity. The alternative is wireless communication. Common uses of wireless data transmission include pagers, cellular telephones, microwave transmissions, communications satellites, mobile data networks, personal communications services, and personal digital assistants (PDAs).

Microwave

Microwave systems are widely used for high-volume, long-distance, point-to-point communication. These systems were first used extensively to transmit very-high-frequency (up to 500 GHz) radio signals at the speed of light in a line-of-sight path between relay stations spaced approximately 30 miles apart (due to the earth's curvature). To minimize line-of-sight problems, microwave antennas were usually placed on top of buildings, towers, and mountain peaks. Long-distance telephone carriers adopted microwave systems because they generally provide about 10 times the data-carrying capacity of a wire without the significant efforts necessary to string or bury wire. Compared to 30 miles of wire, microwave communications can be set up much more quickly (within a day) and at much lower cost.

Satellite

A satellite is a space station that receives microwave signals from an earth-based station, amplifies the signals, and broadcasts the signals back over a wide area to any number of earth-based stations. Transmission to a satellite is an uplink, whereas transmission from a satellite to an earth-based station is a downlink.

A major advance in communications in recent years is the use of communications satellites for digital transmissions. Although the radio frequencies used by satellite data communication transponders are also line-of-sight, the enormous "footprint" of a satellite's coverage area from high altitudes overcomes the limitations of microwave data relay stations. For example, a network of just three evenly spaced communications satellites in stationary "geosynchronous" orbit 22,241 miles above the equator is sufficient to provide global coverage.

The advantages of satellites include the following: The cost of transmission is the same regardless of the distance between the sending and receiving stations within the footprint of a satellite, and cost remains the same regardless of the number of stations receiving that transmission (simultaneous reception). Satellites have the ability to carry very large amounts of data. They can easily cross or span political borders, often with minimal government regulation. Transmission errors in a digital satellite signal occur almost completely at random; thus, statistical methods for error detection and correction can be applied efficiently and reliably. Finally, users can be highly mobile while sending and receiving signals.

The disadvantages of satellites include the following: Any one-way transmission over a satellite link has an inherent propagation delay (approximately one-quarter of a second), which makes the use of satellite links inefficient for some data communications needs (voice communication and "stepping-on" each other's speech).

Global Positioning System

A global positioning system (GPS) is a wireless system that uses satellites to enable users to determine their position anywhere on the earth. GPS is supported by 24 U.S. government satellites that are shared worldwide. Each satellite orbits the earth once in 12 hours, on a precise path at an altitude of 10,900 miles. At any point in time, the exact position of each satellite is known, because the satellite broadcasts its position and a time signal from its on-board atomic clock, accurate to 1-billionth of a second. Receivers also have accurate clocks that are synchronized with those of the satellites. Knowing the speed of signals (186,272 miles per second), it is possible to find the location of any receiving station (latitude and longitude) within an accuracy of 50 feet by triangulation, using the distance of three satellites for the computation. GPS software computes the latitude and longitude and converts it to an electronic map.

GPS equipment has been used extensively for navigation by commercial airlines and ships and for locating trucks. GPS is now also being added to many consumer oriented electronic devices. The first dramatic use of GPS came during the Persian Gulf War, when troops relied on the technology to find their way in the Iraqi desert. GPS also played the key role in targeting for smart bombs. Since then, commercial use has become widespread, including navigation, mapping, and surveying, particularly in remote areas. For example, several car manufacturers (e.g., Toyota, Cadillac) provide built-in GPS navigation systems in their cars. GPSs are also available on cell phones, so you can know where the caller is located.

Radio

Radio electromagnetic data communications do not have to depend on microwave or satellite links, especially for short ranges such as within an office setting. Broadcast radio is a wireless transmission medium that distributes radio signals through the air over both long distances and short distances. Radio is being used increasingly to connect computers and peripheral equipment or computers and local area networks. The greatest advantage of radio for data communications is that no wires need be installed. Radio waves tend to propagate easily through normal office walls. The devices are fairly inexpensive and easy to install. Radio also allows for high data transmission speeds.

Infrared

Infrared (IR) light is light not visible to human eyes that can be modulated or pulsed for conveying information. IR requires a line-of-sight transmission. Many computers and devices have an IrDA (Infrared Data Association) port that enables the transfer of data using infrared light rays. IrDA is a standard defined by the IrDA Consortium. It specifies a way to transfer data wirelessly via infrared radiation. The most common application of infrared light is with television or videocassette recorder remote control units. With computers, infrared transmitters and receivers (or "transceivers") are being used for short-distance connection between computers and peripheral equipment, or between computers and local area networks. Many mobile phones have a built-in infrared (IrDA) port that supports data transfer.

Cellular (Radio) Technology

Mobile telephones, which are being used increasingly for data communications, are based on cellular (radio) technology, which is a form of broadcast radio that is widely used for mobile communications. The basic concept behind this technology is relatively simple: The Federal Communication Commission (FCC) has defined geographic cellular service areas; each area is subdivided into hexagonal cells that fit together like a honeycomb to form the backbone of that area's cellular radio system. Located at the center of each cell is a radio transceiver and a computerized cell-site controller that handles all cell-site control functions.

All the cell sites are connected to a mobile telephone switching office that provides the connections from the cellular system to a wired telephone network and transfers calls from one cell to another as a user travels out of the cell serving one area and into another.

Personal Communication Service

Personal communication service (PCS) uses lower-power, higher-frequency radio waves than does cellular technology. It is a set of technologies used for completely digital cellular devices, including handheld computers, cellular telephones, pagers, and fax machines. The cellular devices have wireless modems, allowing you Internet access and e-mail capabilities. The lower power means that PCS cells are smaller and must be more numerous and closer together. The higher frequency means that PCS devices are effective in many places where cellular telephones are not, such as in tunnels and inside office buildings. PCS telephones need less power, are smaller, and are less expensive than cellular telephones. They also operate at higher, less-crowded frequencies than cellular telephones, meaning that they will have the bandwidth necessary to provide video and multimedia communications.

Personal Digital Assistants

Personal digital assistants (PDAs) are small, handheld computers capable of entirely digital communications transmission. They have built-in wireless telecommunications capabilities. Applications include Internet access, e-mail, fax, electronic scheduler, calendar, and notepad software.

UPS's Delivery Information Acquisition Device (DIAD) is a handheld electronic data collector that UPS drivers use to record and store information, thus helping UPS to keep track of packages and gather delivery information within UPS's nationwide, mobile cellular network. It digitally captures customers' package information, thus enabling UPS to keep accurate delivery records. Drivers insert the DIAD into a DIAD vehicle adapter (DVA) in their delivery vehicles to transmit over UPS's nationwide cellular network for immediate customer use.

Wireless Application Protocol

Wireless Application Protocol (WAP) is a technology that enables wireless transmissions. For example, one popular application that utilizes WAP is i-mode, a wireless portal that enables users to connect to the Internet. Developed by NTT DoCoMo, i-mode provides an always-on connection to the Internet and content sites from popular media outlets, all accessible via color-screen handsets with polyphonic sound. It is charged at actual usage instead of on a prepaid basis. WAP is criticized for browsing with small screens, little compelling content and bad connections at great cost through a browser. Despite these drawbacks, it offers users the ability to make wireless connections to the Internet, which has enormous commercial appeal.

Bluetooth

A relatively new technology for wireless connectivity is called Bluetooth. It is the term used to describe the protocol of a short-range (10meter), frequency-hopping radio link between devices. Bluetooth allows wireless communication between mobile phones, laptops, and other portable devices. Bluetooth technology is currently being built into mobile PCs, mobile telephones, and PDAs.

Bluetooth is the code name for a technology designed to provide an open specification for wireless communication of data and voice. It is based on a low-cost, short-range radio link built into a 9 9 mm microchip, providing protected ad hoc connections for stationary and mobile communication environments. It allows for the replacement of the many existing proprietary cables that connect one device to another with one universal short-range radio link.

Communication Signals

A signal is an electromagnetic or electrical current that carries data from one system or network to another. In electronics, a signal is often a time-varying voltage that is also an electromagnetic wave carrying information, though it can take on other forms, such as current. There are two main types of signals used in electronics: analog and digital signals.

Analog Signal

An analog signal is time-varying and generally bound to a range (e.g. +12V to -12V), but there is an infinite number of values within that continuous range. An analog signal uses a given property of the medium to convey the signal's information, such as electricity moving through a wire. In an electrical signal, the voltage, current, or frequency of the signal may be varied to represent the information. Analog signals are often calculated responses to changes in light, sound, temperature, position, pressure, or other physical phenomena. When plotted on a voltage vs. time graph, an analog signal should produce a smooth and continuous curve. There should not be any discrete value changes.

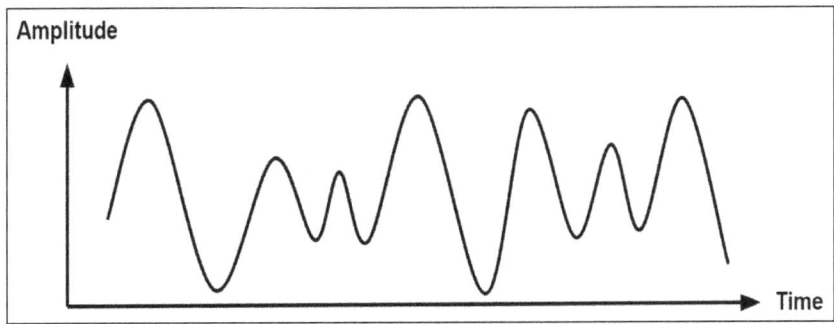

Digital Signal

A digital signal is a signal that represents data as a sequence of discrete values. A digital signal can only take on one value from a finite set of possible values at a given time. With digital signals, the physical quantity representing the information can be many things:

- Variable electric current or voltage.

- Phase or polarization of an electromagnetic field.

- Acoustic pressure.

- The magnetization of a magnetic storage media.

Digital signals are used in all digital electronics, including computing equipment and data transmission devices. When plotted on a voltage vs. time graph, digital signals are one of two values, and are usually between 0V and VCC (usually 1.8V, 3.3V, or 5V).

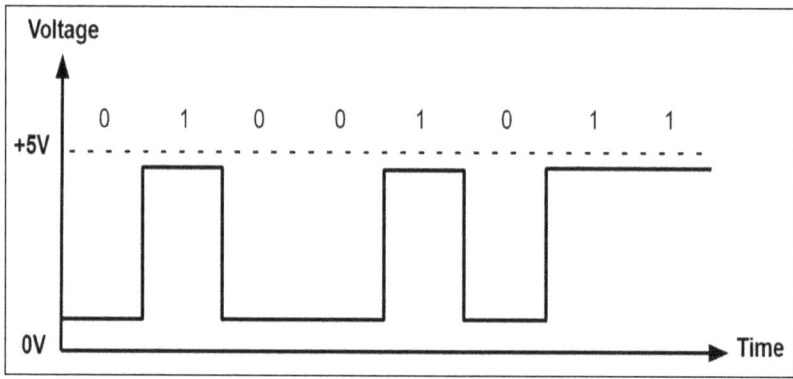

Analog Electronics

Most of the fundamental electronic components — resistors, capacitors, inductors, diodes, transistors, and operational amplifiers (op amps) — are all inherently analog components. Circuits built with a combination of these components are analog circuits.

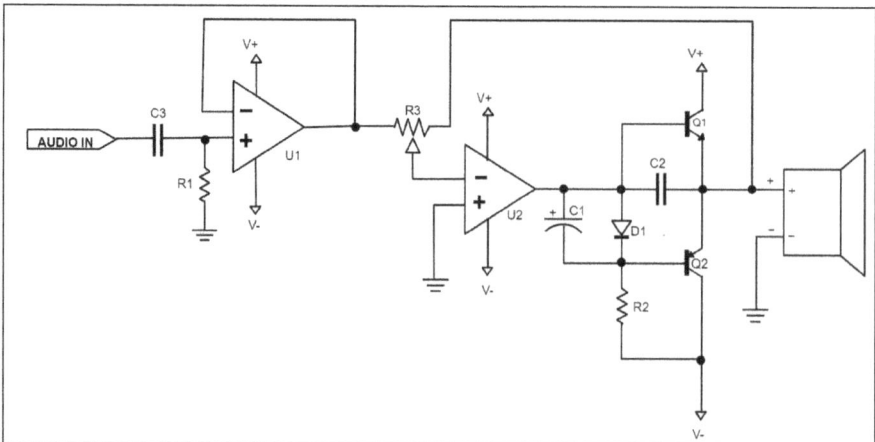

Analog circuits can be complex designs with multiple components, or they can be simple, such as two resistors that form a voltage divider. In general, analog circuits are more difficult to design than digital circuits that accomplish the same task. It would take a designer who is familiar with analog circuits to design an analog radio receiver, or an analog battery charger, since digital components have been adopted to simplify those designs.

Analog circuits are usually more susceptible to noise, with "noise" being any small, undesired variations in voltage. Small changes in the voltage level of an analog signal can produce significant errors when being processed.

Analog signals are commonly used in communication systems that convey voice, data, image, signal, or video information using a continuous signal. There are two basic kinds of analog transmission, which are both based on how they adapt data to combine an input signal with a carrier signal. The two techniques are amplitude modulation and frequency modulation. Amplitude modulation (AM) adjusts the amplitude of the carrier signal. Frequency modulation (FM) adjusts the frequency of the carrier signal. Analog transmission may be achieved via many methods:

- Through a twisted pair or coaxial cable.

- Through an optical fiber cable.

- Through radio.

- Through water.

Much like the human body uses eyes and ears to capture sensory information, analog circuits use these methodologies to interface with the real world, and to accurately capture and process these signals in electronics.

Digital Electronics

Digital circuits implement components such as logic gates or more complex digital ICs. Such ICs are represented by rectangles with pins extending from them.

Digital circuits commonly use a binary scheme. Although data values are represented by just two states (0s and 1s), larger values can be represented by groups of binary bits. For example, in a 1-bit system, a 0 represents a data value of 0, and a 1 represents a data value of 1. However, in a 2-bit system, a 00 represents a 0, a 01 represents a 1, a 10 represents a 2, and a 11 represents a 3. In a 16-bit system, the largest number that can be represented is 216, or 65,536. These groups of bits can be captured either as a sequence of successive bits or a parallel bus. This allows large streams of data to be processed easily.

Unlike analog circuits, most useful digital circuits are synchronous, meaning there is a reference clock to coordinate the operation of the circuit blocks, so they operate in a predictable manner. Analog electronics operate asynchronously, meaning they process the signal as it arrives at the input.

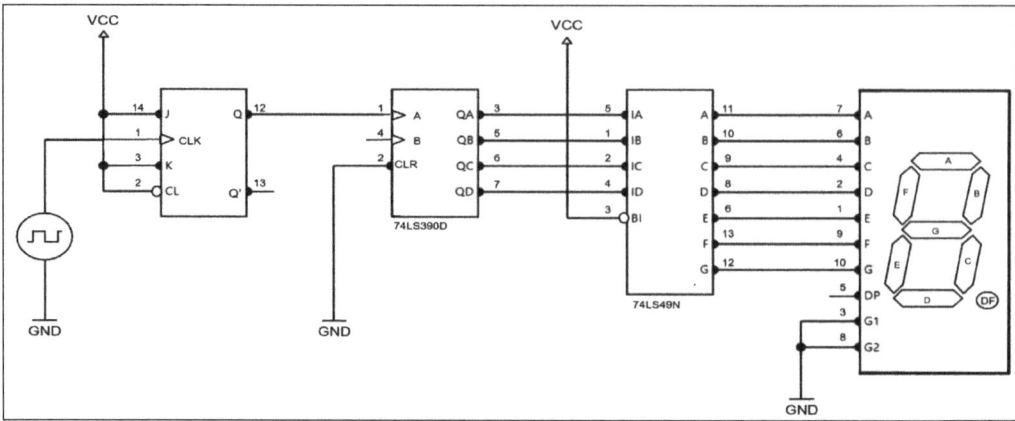

Most digital circuits use a digital processor to manipulate the data. This can be in the form of a simple microcontroller (MCU) or a more complex digital signal processor (DSP), which can filter and manipulate large streams of data such as video. Digital signals are commonly used in communication systems where digital transmission can transfer data over point-to-point or point-to-multipoint transmission channels, such as copper wires, optical fibers, wireless communication media, storage media, or computer buses. The transferrable data is represented as an electromagnetic signal, such as a microwave, radio wave, electrical voltage, or infrared signal.

Analog-to-Digital (ADC) and Digital-to-Analog (DAC) Signal Conversion

Many systems must process both analog and digital signals. It is common in many communications systems to use an analog signal, which acts as an interface for the transmission medium to transmit and receive information.

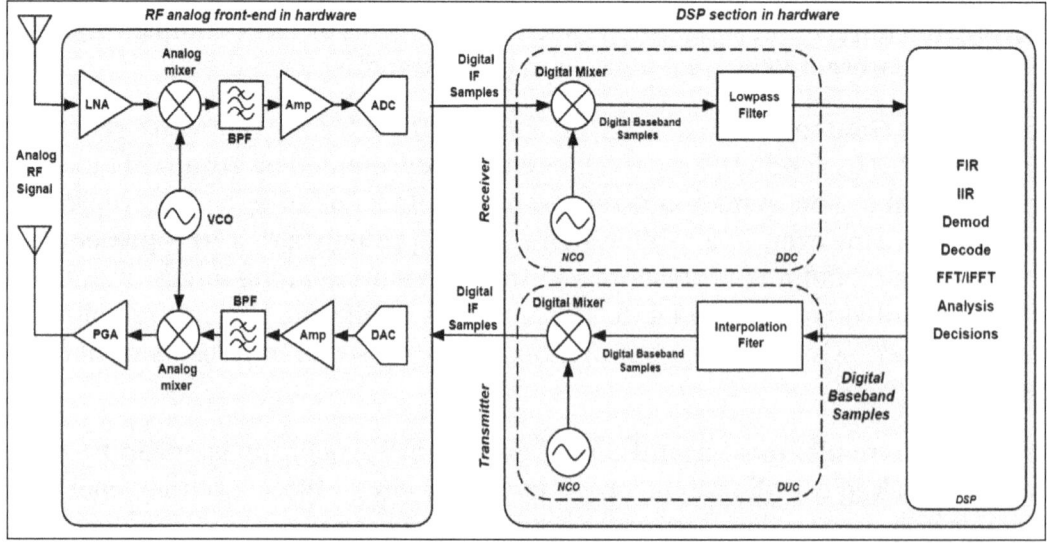

Communication System with Analog and Digital Subsystems.

These analog signals are converted to digital signals, which filter, process, and store the information. Figure shows a common architecture in which the RF analog front-end (AFE) consists of all analog blocks to amplify, filter, and gain the analog signal. Meanwhile, the digital signal processor (DSP) section filters and processes the information. To convert signals from the analog subsystem to the digital subsystem in the receive path (RX), an analog-to-digital converter (ADC) is used. To convert signals from the digital subsystem to the analog subsystem in the transmit path (TX), a digital-to-analog converter (DAC) is used.

A digital signal processor (DSP) is a specialized microprocessor chip that performs digital signal processing operations. DSPs are fabricated on MOSFET integrated circuit chips, and are widely used in audio signal processing, telecommunications, digital image processing, high-definition television products, common consumer electronic devices such as mobile phones, and in many other significant applications.

A DSP is used to measure, filter, or compress continuous real-world analog signals. Dedicated DSPs often have higher power efficiency, making them suitable in portable devices due to their power consumption constraints. A majority of general-purpose microprocessors are also able to execute digital signal processing algorithms.

ADC Operation

Figure below shows ADC operation. The input is the analog signal, which is processed through a sample-hold (S/H) circuit to create an approximated digital representation of the signal. The amplitude no longer has infinite values, and has been "quantized" to discrete values, depending on the resolution of the ADC. An ADC with a higher resolution will have finer step sizes, and will more accurately represent the input analog signal. The last stage of the ADC encodes the digitized signal into a binary stream of bits that represents the amplitude of the analog signal. The digital output can now be processed in the digital domain.

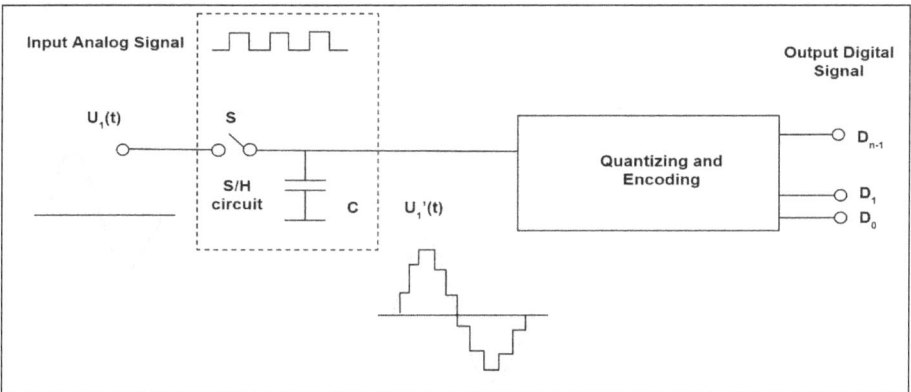

DAC Operation

A DAC provides the reverse operation. The DAC input is a binary stream of data from the digital subsystem, and it outputs a discrete value, which is approximated as an analog signal. As the resolution of the DAC increases, the output signal more closely approximates a true smooth and continuous analog signal. There is usually a post filter in the analog signal chain to further smooth out the waveform.

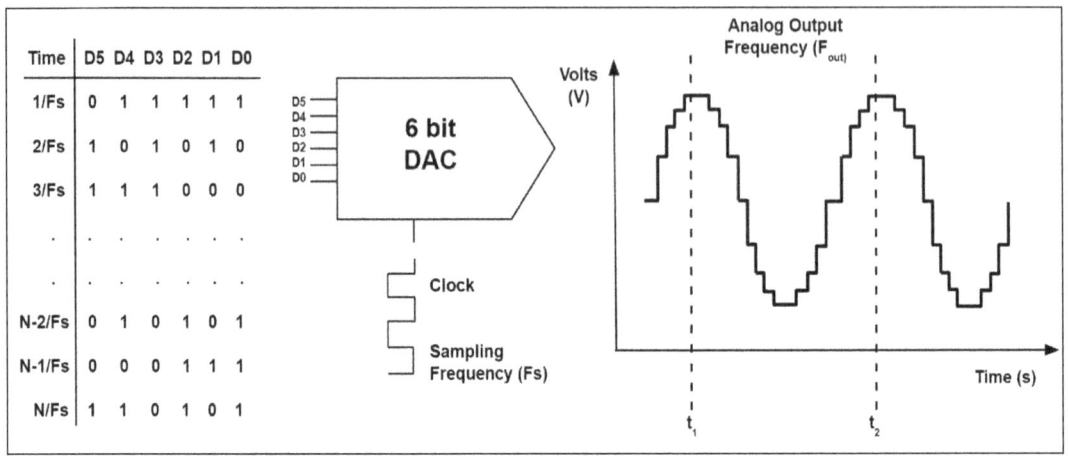

6-Bit DAC for Digital-to-Analog Signal Conversion.

As mentioned before, many systems used today are "mixed signal," meaning they rely on both analog and digital subsystems. These solutions require ADCs and DACs to convert information between the two domains.

Digital Signals vs. Analog Signals: Advantages and Disadvantages

As with most engineering topics, there are pros and cons for both analog and digital signals. The specific application, performance requirements, transmission medium, and operating environment can determine whether analog or digital signaling (or a combination) should be used.

Digital Signals

Advantages to using digital signals, including digital signal processing (DSP) and communication systems, include the following:

- Digital signals can convey information with less noise, distortion, and interference.

- Digital circuits can be reproduced easily in mass quantities at comparatively low costs.

- Digital signal processing is more flexible because DSP operations can be altered using digitally programmable systems.

- Digital signal processing is more secure because digital information can be easily encrypted and compressed.

- Digital systems are more accurate, and the probability of error occurrence can be reduced by employing error detection and correction codes.

- Digital signals can be easily stored on any magnetic media or optical media using semiconductor chips.

- Digital signals can be transmitted over long distances.

Disadvantages to using digital signals, including digital signal processing (DSP) and communication systems, include the following:

- A higher bandwidth is required for digital communication when compared to analog transmission of the same information.

- DSP processes the signal at high speeds, and comprises more top internal hardware resources. This result in higher power dissipation compared to analog signal processing, which includes passive components that consume less energy.

- Digital systems and processing are typically more complex.

Analog Signals

Advantages to using analog signals, including analog signal processing (ASP) and communication systems, include the following:

- Analog signals are easier to process.

- Analog signals best suited for audio and video transmission.

- Analog signals are much higher density, and can present more refined information.

- Analog signals use less bandwidth than digital signals.

- Analog signals provide a more accurate representation of changes in physical phenomena, such as sound, light, temperature, position, or pressure.

- Analog communication systems are less sensitive in terms of electrical tolerance.

Disadvantages to using analog signals, including analog signal processing (ASP) and communication systems, include the following:

- Data transmission at long distances may result in undesirable signal disturbances.

- Analog signals are prone to generation loss.

- Analog signals are subject to noise and distortion, as opposed to digital signals which have much higher immunity.

- Analog signals are generally lower quality signals than digital signals.

Analog and Digital Signals Systems and Applications

Traditional audio and communication systems used analog signals. However, with advances in silicon process technologies, digital signal processing capabilities, encoding algorithms, and encryption requirements — in addition to increases in bandwidth efficiencies — many of these systems have become digital. They are still some applications where analog signals have legacy use or benefits. Most systems that interface to real-world signals (such as sound, light, temperature, and pressure) use an analog interface to capture or transmit the information. A few analog signal applications are listed below:

- Audio recording and reproduction.

- Temperature sensors.

- Image sensors.

- Radio signals.

- Telephones.

- Control systems.

Although many original communication systems used analog signaling (telephones), recent technologies use digital signals because of their advantages with noise immunity, encryption, bandwidth efficiency, and the ability to use repeaters for long-distance transmission. A few digital signal applications are listed below:

- Communication systems (broadband, cellular).

- Networking and data communications.

- Digital interfaces for programmability.

Optical Fiber Communication

Fiber optic communication has revolutionized the telecommunications industry. It has also made its presence widely felt within the data networking community as well. Using fiber optic cable, optical communications have enabled telecommunications links to be made over much greater distances and with much lower levels of loss in the transmission medium and possibly most important of all, fiber optical communications has enabled much higher data rates to be accommodated.

Optical Fiber Communication is the method of communication in which signal is transmitted in the form of light and optical fiber is used as a medium of transmitting those light signal from one place to another. The signal transmitted in optical fiber is converted from the electrical signal into light and at the receiving end; it is converted back into the electrical signal from the light. The data sent can be in the form of audio, video or telemetry data that is to be sent over long distances or over Local Area Networks. Optical fiber communication having good results in long-distance data transfer at high speed, it has been used as an application for various communication purposes.

Working of Fiber Optic Communication

The Optical fiber communication process transmits a signal in the form of light which is first converted into the light from electrical signals and transmitted, and then vice versa happens on the receiving side. This process can be explained using a diagram as shown below:

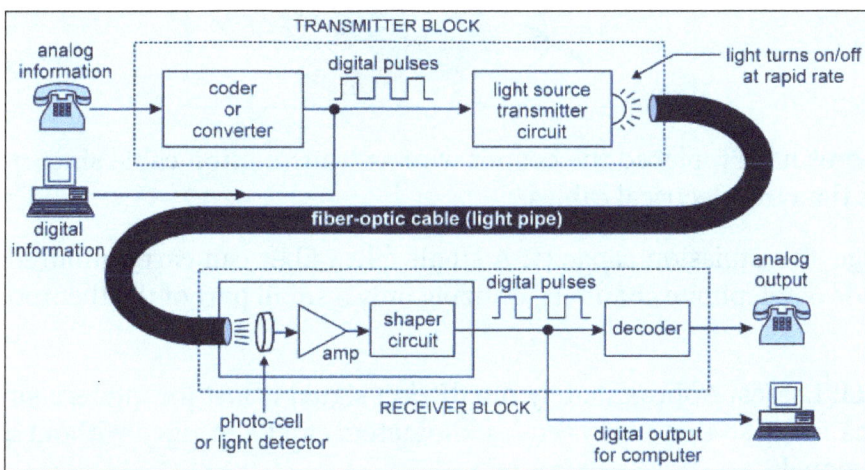

- Transmitter side: On the transmitter side, first if the data is analog, it is sent to a coder or converter circuit which converts the analog signal into digital pulses of 0,1,0,1...(depending on how the data is) and passed through a light source transmitter circuit. And if the input is digital then it is directly sent through the light source transmitter circuit which converts the signal in the form of light waves.

- Optical Fiber Cable: The light waves received from the transmitter circuit to the fiber optic cable is now transmitted from the source location to the destination and received at the receiver block.

- Receiver Side: Now on the receiver side the photocell, also known as the light detector, receives the light waves from the optical fiber cable, amplifies it using the amplifier and converts it into the proper digital signal. Now if the output source is digital then the signal is not changed further and if the output source needs analog signal then the digital pulses are then converted back to an analog signal using the decoder circuit.

The whole process of transmitting an electrical signal from one point to the other by converting it into the light and using Fiber optic cable as transmission source is known as Optical Fiber Communication.

Usage of Fiber

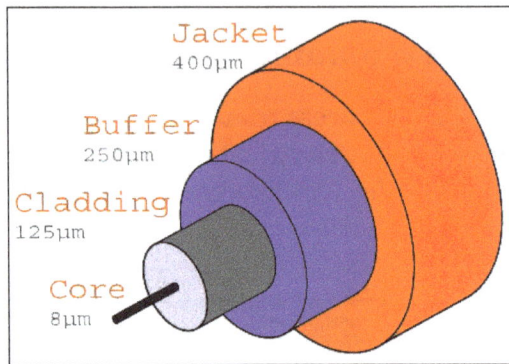

The fiber wires have replaced the copper wire as transmission cable since it has more advantages than the electrical cables:

- Large Transmission capacity: A single silica fiber can carry hundreds of thousands of telephone channels, utilizing only a small part of the theoretical capacity.

- Small Losses: Approximately 0.2 dB/km signal is lost for modern single-mode silica fibers so that many tens of kilometers can be bridged without amplifying the signals.

- Easy Amplification: A large number of channels can be re-amplified in a single fiber amplifier if required for very large transmission distances.

- Low Cost: Due to the huge transmission rate achievable, the cost per transported bit can be extremely low.

- Light Weight: Compared with electrical cables, fiber-optic cables are very light weight.

- No Interference: Fiber-optic cables are immune to problems that arise with electrical cables, such as ground loops or electromagnetic interference (EMI).

The reasons clearly explain that the fiber optic cables are far better than the coaxial copper cables and this is why Fiber optic cables are preferred over the conventions transmission mediums.

Light or Laser light (to be precise) is used for the optical fiber communication because of the reason that the laser light is a single wavelength light source. While the other light signals like sunlight or bulb light have many wavelengths of light and as a result, if used for communication they would produce a beam which is very less powerful and on the other hand, the laser having a single beam would result in a more powerful beam as output.

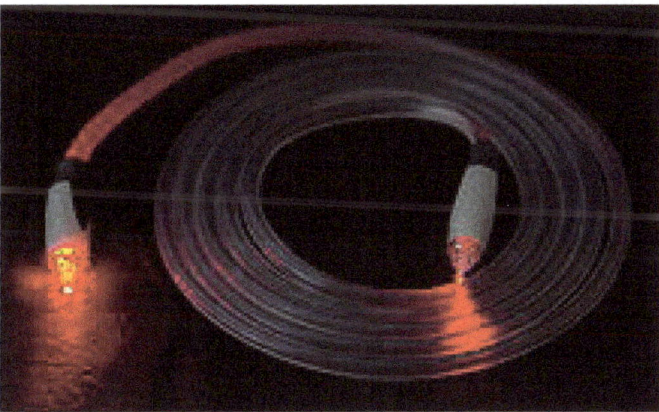

So, Less Dispersion, transmitting more number of signals & consuming less time makes the light a good source for communication.

Characteristics of Fiber Optic Communication

In Optical fiber communication, light is used as a signal which transmitted inside the optical fiber cable. This mode of communication has characteristics which are important to be discussed and makes it a good mode of communication.

- Bandwidth: Single laser light dispersion means that a good amount of signal can be transmitted (Information being transferred in bits) per second which results in high bandwidth for long distances.

- Smaller diameter: The diameter of Optical fiber cable is about 300 micrometers in diameter.

- Light-weight: The Optical fiber cable is light in weight compared to the copper cable.

- Long-distance signal transmission: Since the laser light doesn't disperse, it can be easily transmitted over long distances.

- Low attenuation: The fiber is made of glass and laser is traveling through it, the signal transmitted has only 0.2 dB/km loss.

- Transmission security: Optical encryption and no presence of the electromagnetic signal make the data secure over optical fiber.

Applications of Optical Fiber

Optical fiber communication is mainly used in the telecommunication industry which uses the optical fiber for:

- Telephone Signals transmission.

- Internet Communication.

- Cable Television Signal transmission.

Apart from it, optical fiber nowadays, is used everywhere in homes, industries, offices for long distance as well as for small distance communication.

Optical Fiber Impact on IoT (Internet of Things)

The Fiber Optics Communication will have a great impact on IOT and these things listed will explain to you how IOT would require Fiber Optics:

- Fast Transmission Media: The future will be IOT and all of our devices and things will be connected to the internet, which needs good communication and high speed. The only transmission media that supports such a requirement is Optical Fiber. The future needs IOT and IOT need Optical fiber for best communication that could help reach Wireless data speed up to 100 Gbps speed, making communications and large size data transfer in seconds.

- Data Security: Security in IoT is the main concern when we think of large amount of data to be transferred between billions of devices connected together. Hacking of data from communication media is possible unless it is Optical fiber. The optical fibers are very difficult to hack and hacking them without being detected is like next to impossible. So again, an optical fiber can help secure the data and transfer it at very high speed.

- No data loss due to interference: The optical fiber cables can be installed anywhere (even underwater or at high-temperature areas) and don't have any electromagnetic interference resulting in no data loss due to interference.

References

- How-optical-fiber-communication-works-and-why-it-is-used-in-high-speed-communication: circuitdigest.com, Retrieved 28, May 2020

- Optical-fibre-telecommunications-basics, fibre-optics, connectivity: electronics-notes.com, Retrieved 13, Jan 2020

- Analog-vs-digital-signal: monolithicpower.com, Retrieved 06, June 2020

- Basic-elements-of-a-telecommunication-system: byteintobigdata.in, Retrieved 25, August 2020

- What-is-digital-terrestrial-tv: freeview.co.uk, Retrieved 06, April 2020
- Pros-and-cons-of-digital-telephony: eukhost.com, Retrieved 04, Feb 2020
- What-is-telephony-3426734: lifewire.com, Retrieved 17, May 2020

2

Telecommunication Network and its Types

A telecommunications network is a transmission system consisting of a collection of interconnected nodes, links and intermediate nodes that allow the transmission of information. There are many types of telecommunications network employed globally like the wide area networks, local area networks, virtual private networks, client/server networks, etc. This chapter has been carefully written to provide the reader with a better understanding of the subject matter.

A telecommunications network is a collection of terminal nodes, links and any intermediate nodes which are connected so as to enable telecommunication between the terminals. The transmission links connect the nodes together. The nodes use circuit switching, message switching or packet switching to pass the signal through the correct links and nodes to reach the correct destination terminal.

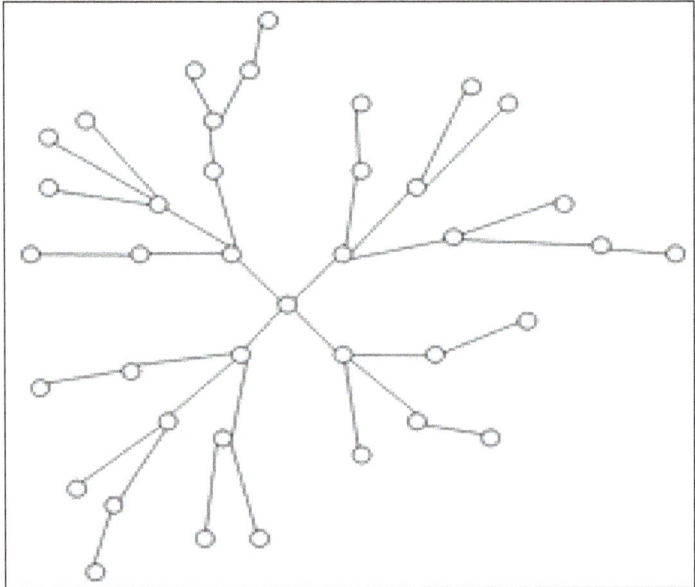

Example of how nodes may be interconnected with links to form a telecommunications network. This example is tree-like but many networks have loops.

Each terminal in the network usually has a unique address so messages or connections can be routed to the correct recipients. The collection of addresses in the network is called the address space. Examples of telecommunications networks are:

• Computer networks.

- The Internet.

- The telephone network.

- The global Telex network.

- The aeronautical ACARS network.

Benefits of Telecommunications and Networking

Telecommunications can greatly increase and expand resources to all types of people. For example, businesses need a greater telecommunications network if they plan to expand their company. With Internet, computer, and telephone networks, businesses can allocate their resources efficiently. These core types of networks will be discussed below:

- Computer Network: A computer network consists of computers and devices connected to one another. Information can be transferred from one device to the next. For example, an office filled with computers can share files together on each separate device. Computer networks can range from a local network area to a wide area network. The difference between the types of networks is the size. These types of computer networks work at certain speeds, also known as broadband. The Internet network can connect computer worldwide.

- Internet Network: Access to the network allows users to use many resources. Over time the Internet network will replace books. This will enable users to discover information almost instantly and apply concepts to different situations. The Internet can be used for recreational, governmental, educational, and other purposes. Businesses in particular use the Internet network for research or to service customers and clients.

- Telephone Network: The telephone network connects people to one another. This network can be used in a variety of ways. Many businesses use the telephone network to route calls and/or service their customers. Some businesses use a telephone network on a greater scale through a private branch exchange. It is a system where a specific business focuses on routing and servicing calls for another business. Majority of the time, the telephone network is used around the world for recreational purposes.

In general, every telecommunications network conceptually consists of three parts, or planes (so called because they can be thought of as being, and often are, separate overlay networks):

- The control plane carries control information (also known as signaling).

- The data plane or user plane or bearer plane carries the network's user's traffic.

- The management plane carries the operations and administration traffic required for network management.

Example of the TCP/IP Data Network

The data network is used extensively throughout the world to connect individuals and organizations. Data networks can be connected to allow users seamless access to resources that are hosted outside of the particular provider they are connected to. The Internet is the best example of many data networks from different organizations all operating under a single address space.

Terminals attached to TCP/IP networks are addressed using IP addresses. There are different types of IP address, but the most common is IP Version 4. Each unique address consists of 4 integers between 0 and 255, usually separated by dots when written down, e.g. 82.131.34.56.

TCP/IP are the fundamental protocols that provide the control and routing of messages across the data network. There are many different network structures that TCP/IP can be used across to efficiently route messages, for example:

- Wide area networks (WAN).

- Metropolitan area networks (MAN).

- Local area networks (LAN).

- Internet area networks (IAN).

- Campus area networks (CAN).

- Virtual private networks (VPN).

There are three features that differentiate MANs from LANs or WANs:

- The area of the network size is between LANs and WANs. The MAN will have a physical area between 5 and 50 km in diameter.

- MANs do not generally belong to a single organization. The equipment that interconnects the network, the links, and the MAN itself are often owned by an association or a network provider that provides or leases the service to others.

- A MAN is a means for sharing resources at high speeds within the network. It often provides connections to WAN networks for access to resources outside the scope of the MAN.

Optical Transport Network (OTN)

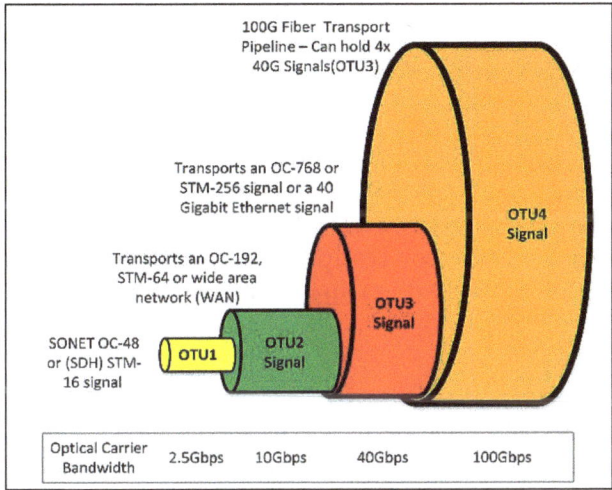

Optical Transport Carrier Specification.

Optical Transport Network (OTN) is a large complex network of server hubs at different locations on ground, connected by Optical fiber cable or optical network carrier, to transport data across different nodes. The server hubs are also known as head-ends, nodes or simply, sites. OTNs are the backbone of Internet Service Providers and are often daisy chained and cross connected to provide network redundancy. Such a setup facilitates uninterrupted services and fail-over capabilities during maintenance windows, equipment failure or in case of accidents. The devices used to transport data are known as network transport equipment. Some of the widely used equipment is manufactured by:

- Alcatel Lucent: AL7510, AL7750.

- Nortel Networks Corp. (acquired by Ciena Corp): Optera Metro series – OM4500, OM6500.

- Fujitsu Ltd: FlashWave series FW4500, FW7500, FW9500.

The capacity of a network is mainly dependent on the type of signaling scheme employed on transmitting and receiving end. In the earlier days, a single wavelength light beam was used to transmit data, which limited the bandwidth to the maximum operating frequency of the transmitting and receiving end equipment. With the application of wavelength division multiplexing (WDM), the bandwidth of OTN has risen up to 100Gbit/s (OTU4 Signal), by emitting light beams of different wavelengths. Lately, AT&T, Verizon, and Rogers Communication have been able to employ these 100G "pipes" in their metro network. Large field areas are mostly serviced by 40G pipes (OC192/STM-64).

A 40G pipe can carry 40 different channels as a result of Dense Wave Division

Multiplexing (DWDM) transmission. Each node in the network is able to access different channels, but is mostly tuned to a few channels. The data from a channel can be dropped to the node or new data can be added to the node using Re-configurable Optic Add Drop Mux (ROADM) that uses Wavelength Selective Switching (WSS) to extract and infuse a configured frequency. This eliminates the need to convert all the channels to electric signals, extract the required channels, and convert the rest back to optical into the OTN. Thus ROADM systems are fast, less expensive and can be configured to access any channel in the OTN pipe. The extracted channels at a site are connected to local devices through muxponder or tranponder cards that can split or combine 40G channels to 4x 10G channels or 8x 2.5G channels.

Wide Area Network

A WAN is a network that uses various links—private lines, Multiprotocol Label Switching (MPLS), virtual private networks (VPNs), wireless (cellular), the Internet—to connect smaller metropolitan and campus networks in diverse locations into a single, distributed network. The sites they connect could be a few miles apart or halfway around the globe. In an enterprise, the purposes of a WAN could include connecting branch offices or even individual remote workers with headquarters or the data center in order to share corporate resources and communications.

WANs vs. LANs: What's the Difference?

A WAN is often contrasted with a local area network or LAN. LANs are networks generally limited to a single building or small campus. They're private to a single organization or even person and can be created with relatively inexpensive equipment. Your home Wi-Fi network is a LAN.

The technologies and protocols that make LANs easy to set up don't scale beyond a certain limited distance or to truly massive numbers of endpoints. Dealing with those scales is the purpose of a WAN: connecting one or more LANs. The networking technologies and protocols WANs use to transmit information are different from those used within LANs.

The Internet is, strictly speaking, a WAN. However, when we talk about WANs, we usually mean private or semi-private networks combining far-flung LANs. Branch offices in different cities might share private internal corporate resources over a WAN, for instance.

WAN Architecture

While LANs are usually maintained by an organization's own IT staff, WANs are often at least in part reliant on physical connections provided by telecommunications carriers. Decisions on what kind of connections or communications protocols to use and how to deploy them will guide the creation of your WAN architecture.

WAN Protocols

Let's start with WAN protocols—the rule sets that define network communication over a WAN. One of the earliest protocols used to deliver WAN traffic is X.25, which uses packet-switching exchanges (PSE) for the hardware that drops traffic onto the wires connecting sites. It includes standard-sized packets delivered in order and includes error correction. The physical links include leased lines, dialup telephone services or Integrated Services Digital Network (ISDN) connections. It's not used much anymore.

Frame relay is a successor to X.25. Frame relay cuts data into different-sized frames and leaves error correction and retransmission of missing packets up to the endpoints. These differences speed up the overall data rate. In addition, Frame Relay relies less on dedicated connections to create meshed networks, meaning fewer physical circuits, hence saving companies money. Frame relay, while once extremely popular, has become less so.

Asynchronous Transfer Mode (ATM) is similar to frame relay with one big difference: data is broken into standard-sized packets called cells. Cells make it possible to blend different classes of traffic onto a single physical circuit and more readily guarantee qualities of service. The downside of ATM is that because it uses relatively small cells, the headers eat up a relatively large percentage of the total contents of transmissions. Therefore, ATM's overall use of bandwidth is less efficient than that of frame relay. ATM also has fallen out of favor with business customers.

Today, multi-protocol label switching is used to carry much corporate data across WAN links. Within an MPLS network, brief header segments called labels allow MPLS routers to decide quickly where to forward packets and to treat them with the class of service indicated by the labels. This makes it possible to run different protocols within MPLS packets while giving different applications appropriate priority as traffic travels between sites. Internet protocol (IP), which became more ubiquitous in the 1990s, is one protocol commonly carried within MPLS.

Types of WAN Connections

All these protocols operate over different kinds of network connections. Initially, WANs were built with meshed webs of private lines bought from telecommunications carriers, but WAN architectures have advanced to include packet-switched services such as frame relay and ATM as well as MPLS. With these services, a single connection to a site can be connected to many others via switching within service-provider networks. These types of connections provide direct and largely private means of communication for your various LANs. That gets you speed and security—but it isn't cheap. For certain types of traffic, the Internet can also be woven into the mix to provide less expensive WAN connections.

What is Tunneling

WAN connections that operate over the Internet or some other public network generally use a technique known as tunneling. In a tunneled connection, the private-network data and protocol information are encrypted and encapsulated in IP packets that are routed over the open internet. When those packets arrive at the destination LAN, the IP headers are stripped away, the payload is decrypted and private-networking features come back into play. From the perspective of the LAN users at either end, the packets behave as if they're travelling over a private WAN. The name for the technique comes from the metaphorical tunnel that the private packets travel through.

What is a VPN

The most common tunnel is the virtual private network (VPN). VPN connections encrypt data in order to keep it private as it travels over public networks. VPNs are frequently used to allow home office workers to connect to private corporate WANs. A VPN user's Internet traffic is routed through the WAN network they're connected to, which can give them an IP address that doesn't reflect their real physical location; this makes VPNs a favored tool for streaming content that may be restricted by geography.

SD-WANs

WANs today may use multiple types of connections and protocols simultaneously, which obviously adds many layers of complexity. As a result, the use of software-defined technology to manage WANs is gaining momentum. Software-defined WAN (SD-WAN) takes software-defined concepts, especially the decoupling of the control plane from the data plane, and brings it to the WAN.

SD-WAN uses software to monitor the performance of a mix of WAN connections—MPLS, dedicated circuits, the Internet—and to choose the most appropriate connection for each traffic type. So teleconferencing might run over a dedicated circuit, but email might use the internet. In making its decisions, SD-WAN software takes into account how well each link is performing at the moment, the cost of each connection and the needs of each application.

Initially SD-WAN aimed at creating hybrid WANs and using policies to mix MPLS and internet connections in order to improve efficiency and lower costs. The next phase will improve management and monitoring and provide better security, according to Lee Doyle of Doyle Research. SD-WAN connections proved invaluable as office workers scattered to their homes during the 2020 coronavirus pandemic, and the market is expected to increase by 168% by 2024, according to the Dell'Oro Group.

A subset of SD-WAN called SD-Branch is helping reduce the need for hardware within branch offices. Offerings from big vendors including Aruba and Juniper can replace many physical devices with software running on off-the-shelf servers. Mobile backup across SD-WAN can provide a failover for broadband connections as wireless WAN technology (4G, LTE, 5G, etc.) costs decrease.

WAN Management and Optimization

Because data transmission is still reliant on the rules of physics, the greater the distance between two devices, the longer it will take for data to travel between them. The greater the distance, the greater the delay. Network congestion and dropped packets can also introduce performance problems.

Some of this can be addressed using WAN optimization, which makes data transmissions more efficient. This is important because WAN links can be expensive, so technologies have sprung up that reduce the amount of traffic crossing WAN links and ensure that it arrives efficiently. These optimization methods include abbreviating redundant data (known as de-duplication), compression, and caching (putting frequently used data closer to the end user).

Traffic can be shaped to give time-sensitive applications such as VoIP a higher priority over other, less urgent traffic such as email, which in turn helps improve the overall WAN performance. This can be formalized into quality-of-service settings that define classes of traffic by the priority each class receives relative to others, the type of WAN connection that each traffic type will travel, and the bandwidth that each receives. Once a separate category, WAN optimization is being absorbed by SD-WAN.

WANs have been around since the early days of computing networks. WANs were based on circuit-switched telephone lines and modems but now connectivity options also include leased lines, wireless, MPLS, broadband internet, and satellite.

As technologies changed, so did transmission rates. The early days of 2400bps modems evolved to 40Gbps and 100Gbps connectivity today. These speed increases have allowed more devices to connect to networks, enabling the explosion of connected computers, phones, tablets and smaller Internet of Things devices.

In addition, speed improvements have allowed applications to utilize larger amounts of bandwidth that can travel across WANs at super-high speed. This has allowed enterprises to implement applications such as videoconferencing and large-file data backup. Nobody would have considered conducting a videoconference across a 28kbps modem, but now workers can sit at home and participate in global company meetings via video.

Many WAN links are supplied via carrier services in which customers' traffic rides over facilities shared by other customers. Customers can also buy dedicated links that nail up circuits point-to-point and are used for just one customer's traffic. These are typically used for top-priority or delay-sensitive applications that have high-bandwidth needs such as videoconferencing.

WAN Security

Traffic between WAN sites may be protected by virtual private networks (VPN) that overlay security on the underlying physical network, including authentication, encryption, confidentiality and non-repudiation. In general, security is a crucial part of any WAN rollout, because a WAN connection represents a potential vulnerability that an attacker could use to gain access to a private network.

For instance, a branch office without a full-time infosec staffer might be lax in its cyber security practices. As a result, a hacker who breached the network at the branch could go on to access the main corporate WAN, including valuable assets that would have been otherwise impregnable. In addition to networking features, many SD-WAN offerings provide security services as well, which need to be kept top of mind during deployment.

Interplanetary Internet

WAN technologies aren't just limited to Earth. NASA and other space agencies are working to create a reliable "interplanetary internet," which aims to transmit test messages between the International Space Station and ground stations. The Disruption Tolerant Networking (DTN) program is the first step in providing an Internet-like structure for communications between space-based devices, including communicating between the Earth and Moon, or other planets. But barring any dramatic breakthrough in physics, network speeds would likely top out at the speed of light.

Types of WAN Technologies

- Packet switching: Packet switching is a method of data transmission in which a message is broken into several parts, called packets that are sent independently, in triplicate, over whatever route is optimum for each packet and reassembled at the destination. Each packet contains a piece part, called the payload, and an identifying header that includes destination and reassembly information. The packets are sent in triplicate to check for packet corruption. Every packet is verified in a process that compares and confirms that at least two copies match. When verification fails, a request is made for the packet to be re-sent.

- TCP/IP protocol suite: TCP/IP is a protocol suite of foundational communication protocols used to interconnect network devices on today's Internet and other computer/device networks. TCP/IP stands for Transmission Control Protocol/Internet Protocol.

- Router: A router is a networking device typically used to interconnect LANs to form a wide area network (WAN) and as such is referred to as a WAN device. IP routers use IP addresses to determine where to forward packets. An IP address is a numeric label assigned to each connected network device.

- Overlay network: An overlay network is a data communications technique in which software is used to create virtual networks on top of another network, typically a hardware and cabling infrastructure. This is often done to support applications or security capabilities not available on the underlying network.

- Packet over SONET/SDH (PoS): Packet over SONET is a communication protocol used primarily for WAN transport. It defines how point-to-point links communicate when using optical fiber and SONET (Synchronous Optical Network) or SDH (Synchronous Digital Hierarchy) communication protocols.

- Multiprotocol Label Switching (MPLS): MPLS is a network routing-optimization technique. It directs data from one node to the next using short path labels rather than long network addresses, to avoid time-consuming table lookups.

- ATM: ATM (Asynchronous Transfer Mode) is a switching technique common in early data networks, which has been largely superseded by IP-based technologies. ATM uses asynchronous time-division multiplexing to encode data into small, fixed-sized cells. By contrast, today's IP-based Ethernet technology uses variable packet sizes for data.

- Frame Relay: Frame Relay is a technology for transmitting data between LANs or endpoints of a WAN. It specifies the physical and data-link layers of digital telecommunications channels using a packet switching methodology.

  Frame Relay packages data in frames and sends it through a shared Frame Relay network. Each frame contains all necessary information for routing it to its destination. Frame Relay's original purpose was to transport data across telecom carriers' ISDN infrastructure, but it's used today in many other networking contexts.

IPsec VPN

IPsec VPN securely connects all of your sites on the same private network using Internet connectivity as the data communications network. This type of VPN is deployed between a security appliance and firewall at each location, ensuring a secure IPsec tunnel between sites. The LAN sits behind these security devices and software isn't required on laptops, desktops, or servers to enable VPN connectivity between locations. VPN network topologies are available in a hub and spoke or meshed configuration.

The main benefits of these types of data networking services are cost, the ability to use existing Internet connectivity for data transport, and easy integration of remote users with VPN software. IPsec VPN connectivity does have its flaws in that the quality of service (QoS) is not consistent due to Internet network congestion or poor performance. Also, there is increased potential for network downtime if only using one

Internet connection with no failover connectivity. IPsec VPN networks are a good choice for businesses with limited IT budgets, many remote users, or basic applications and uptime requirements.

Software-Defined WAN (SD-WAN)

Software-Defined WAN (SD-WAN) is an emerging type of WAN technology. Software-Defined Networking (SDN) is used to automatically determine the best routes to and from locations over Internet connections and private data networks. SD-WAN creates tunnels that are transport-agnostic, so you can use Internet connections like DSL, cable, wireless, shared fiber, or dedicated connectivity. Businesses can also keep existing private data network services (MPLS, EPL, EVPL, VPLS, etc.) in addition to regular Internet connectivity. This helps to improve SD-WAN network performance and reliability.

SD-WAN use multiple tunnels to increase and optimize WAN bandwidth between different types of WAN technologies, a big advantage over traditional IPsec VPNs. This ensures applications have the highest QoS, increased WAN speeds, as well as additional network redundancy and failover. SD-WAN centralize network control and traffic management over these links through a centralized controller or orchestrator.

VPN security is layered on top, while SDN software enables IT staff to remotely manage network edge devices and applications more easily. SD-WAN is a good option for businesses of all sizes and needs. Enabling data networking over low-cost Internet connectivity, as well as more expensive dedicated WAN links, is a big plus. Increased reliability, performance, network agility are all key features of SD-WAN service, along with a competitive price point.

Metro Ethernet

Metro Ethernet is a point-to-point Ethernet data networking service connecting locations within a metropolitan area (MAN). Ethernet over Synchronous Optical Network (SONET) technology is used for secure point to point WAN connectivity. Circuit speeds typically range from 10 Mbps to 10 Gbps, with 100Gbps available in some metropolitan areas.

Provider networks are Layer 2, so you have control over addressing and routing. Metro Ethernet service is ideal for businesses with two or more locations in a metro area that need high bandwidth connectivity with QoS requirements. In most cases, average costs for Metro Ethernet service tend to be low due to minimal distances and limited network infrastructure used to provide service.

Ethernet Private Line (EPL)

Ethernet Private Line service (EPL) provides dedicated point-to-point Ethernet network connectivity between two or more locations. Like Metro Ethernet, Ethernet over Synchronous Optical Network (SONET) is the type of WAN technology used. EPL circuits provide a reliable data networking service for customers with high bandwidth and low latency needs. A key component of EPL service is network resiliency and performance through SONET protection (network reroute). Making this data networking service for the most mission-critical applications.

Being a Layer 2 network, addressing and routing is customer controlled. Ethernet Private Line is available from 10 Mbps to 10Gbps, with 100Gbps available in some locations. EPL is one of the more expensive types of WAN technologies due to distance-sensitive pricing and dedicated network infrastructure used.

MPLS VPN

MPLS VPN is a virtual private network built on top of a provider's Multiprotocol Label Switching Network to provide Layer 2 or Layer 3 VPN data networking services. Multiprotocol and tagging capabilities of MPLS connect remote sites into a common type of data communication network. the configurations available include site to site, multipoint, and meshed networks. Customer data is partitioned from each other, keeping it private across the provider's infrastructure. Data partitioning is created using MPLS tags rather than encryption.

MPLS is different from other VPN data networking services due to the fact that you can prioritize traffic types over the MPLS providers' network. This allows control over application performance (low to high QoS). MPLS circuit speeds typically range from 10 Mbps to 10Gbps, with costs similar to dedicated Internet connectivity. MPLS networks are the current industry standard for a private data networking service, due to its superior performance, reliability, flexibility, and competitive pricing.

Ethernet Virtual Private Line (EVPL)

Ethernet Virtual Private Line (EVPL) or E-Line provides point-to-multipoint connectivity over a provider's MPLS network. EVPL uses Ethernet Virtual Connections (EVCs) to connect multiple locations together, as well as multiple services on a single User-to-Network Interface (UNI) at the host or hub site.

EVPL is a Layer 2 data networking service utilizing MPLS tags and supporting multiple classes of service (CoS) for low to high QoS applications. Ethernet Virtual Private Lines are available from 10 Mbps to 10 Gbps. EVPL is ideal for customers looking for a reliable type of data communications network for a hub site to multiple remote locations. Due to shared network infrastructure and limited distance costs, EVPL pricing is not as expensive as EPL.

Virtual Private LAN Service (VPLS)

Virtual Private LAN Services (VPLS) or E-LAN is a data network service for multiple sites in a single bridged domain over a provider managed MPLS network. All sites on a VPLS network will appear to be on the same LAN, regardless of the location. Like

EVPL, it is a Layer 2 type of data communications network that utilizes MPLS tags. VPLS also supports multiple classes of service (CoS) for low to high QoS application needs. Multiple types of WAN technologies (MPLS VPN, Internet, and EVPL) are supported on a single port and circuit.

Routing and management of the VPLS network can be done by the customer or provider. VPLS networks offer the ability of a meshed network configures (any to any), so all sites can communicate with each other, increasing network continuity. VPLS is ideal for customers looking for reliable connectivity for a hub site to many remote locations. VPLS speeds are usually 10Mbps to 10Gbps with pricing comparable to EVPL and MPLS data networking services.

Wavelengths

Wavelength Service is an optical data networking solution for customers requiring very large dedicated point-to-point data connections. This is ideal for business continuity, data center replication, backup solutions, streaming media, or very large data transfers. Applications that require low latency and high-speed connectivity are ideal for this type of WAN technology.

Speeds typically available are 2.5Gbps, 10Gbps, 40Gbps, 100Gbps delivered as an optical handoff. Wavelength service is provisioned over a Dense Wave Division Multiplexing (DWDM) network, providing full Layer 2 transparency and management. Unprotected and protected network reroute is available to ensure the resiliency of data network connectivity. Wavelength service has the lowest per Gbps cost of all data networking services, but an overall higher price point due to large bandwidth sizes. This type of WAN technology is typically a wholesale application used by ISP's, telecoms, data centers, media, and big tech.

Metropolitan Area Networks

A Metropolitan Area Network is a computer network based in a metropolitan or city area, through a connection of multiple LAN accessed by point to point connections. By uniting smaller LAN networks, MANs have the ability to extend to a geographical area of 5 to 50 kms, sufficient for the network requirements of a city. This interconnection of networks across the city may also end up facilitating connections to WAN.

MANs are particularly useful for the purpose of sharing resources specific to regions, assisting in quicker and segmented flow of data and information. Furthermore, MANs can connect multiple computers and networks to serve as an Internet Service Provider, because of which the internet service provided by a telephone company falls under the category of a MAN connection. A Metropolitan Area Network is smaller than the area covered through WANs, but it is much bigger than the connectivity capabilities of a LAN. The wide acceptability and applicability of MAN is apparent from its deployment in areas of Geneva, Switzerland; in London, England and also Lodz, Poland. Recently, MANs have been started to be deployed in wireless medium, further improving its ease of accessibility.

Working Methodology

The working mechanism of a MAN is quite similar to that of an Internet Service Provider; however, it is not owned by a single organization. Just like a Wide Area Network, a MAN too delivers a joint and shared network connection to all of its users based on a data link layer. This data link layer is classified as a Layer 2 of the OSI model, an abbreviation for Open Systems Interconnection.

Additionally, Distributed Queue Dual Bus (DQDB) is the standard set by the IEEE, an organization responsible for overlooking the activities of electrical and electronic engineers. Using this standard set by the Institute of Electrical and Electronics Engineers, a MAN expands over 5km-50km range, which is more than enough to cover a city.

The primary goal of a MAN is to establish a connection between geographically separated LANs, meaning initially a MAN seeks to form a communication link between two independent LAN nodes. Established using optical fibers cables, a MAN utilizes routers and switches along with a modem. Just to refresh the memory, a switch is responsible for filtering data inflowing in the shape of frames. The switch lies as one of the fundamental components as it is actively responsible for dual tasks. At one end, as mentioned before, it filters data and on the other end it manages the connection. The router on the other hand plays a role in facilitating the establishment of a network connection by assisting a data packet through directing it to take the appropriate path, hence keeping a vigilant eye on the entire data transfer process.

Types of MAN Technologies

FDDI

Fiber Distribution data interface is a standard for data transfer in context to a LAN and it can assist in transmitting the data of thousands of users. It uses optical fiber for its primary infrastructure; hence the name fiber distribution data interface.

SMDS

Switched multi-megabit data service allows the transfer of data through a connectionless service. What does a connectionless service imply? It is the state of data transfer when the information and data is stored in a head and then it reaches its specific destination in an independent manner. Data can be transmitted over huge geographical distances through the usage of datagrams created by SMDS, which are packets of data of an unreliable data service provider.

ATM

Asynchronous Transfer Mode is the most frequently used MAN technology. ATM is a digital transfer technology, developed in the 1980s to transfer real time data over an individual network. In ATM, the data is stored in specific fixed sized packets transferring overtime; much similar to how a cell relay system is operated. It consists of circuit switching and packet switching, further facilitating the movement of real time data.

Although these technologies may seem complex at first sight to any network enthusiast or a potential networking professional, with the adequate networking training, all of this turns out to be easily and comprehensively understandable.

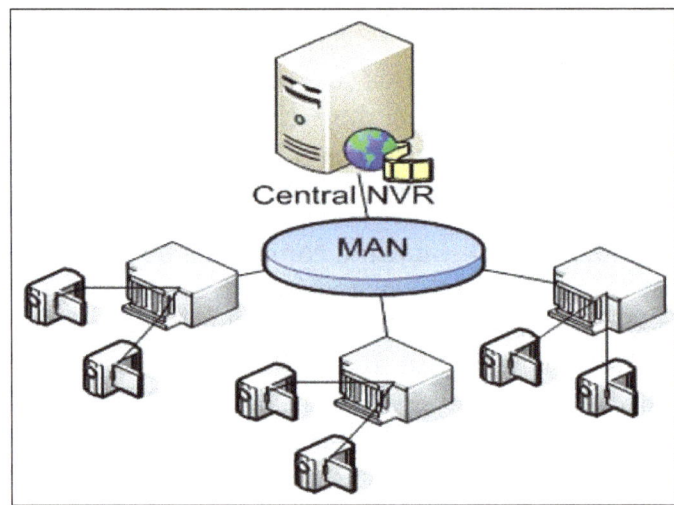

Benefits and Disadvantages of Metropolitan Area Networks

Metropolitan area networks, according to their design, may deliver some benefits, such as firewall centralization. Having a more central gatekeeping point for the Internet can cut down on malware and other threats. A MAN may also provide more efficient forms of administration and data entry. Cities implementing modern MANs may also have complementary technology implementations, such as Internet Exchange points.

However, MANs also carry some particular disadvantages. One of these is buy-in – the idea that there must be some coherent push to adopt a metropolitan area network, and accommodate its use, which is where some of these implementations run into problems. One barrier to implementing MANs is that ISPs often have objections to metropolitan area networks.

The reason is because an effective metropolitan area network makes it more difficult for ISPs to collect the fees that they get from managing local area networks in that covered region or area. It's not hard to find evidence of this, if you look.

For example, a series of reports from Ars Technica from 2010 to 2014 provide ample evidence of ISPs working against MAN implementation, some citing such efforts in 20 U.S. states. In general, we have seen the pushback against MANs hobble efforts to put these systems in place in many U.S. communities. The frontier of networking clashes with a profit motive. This is one central factor in the future of this type of municipal or civic networking option.

Local Area Network

A Local Area Network (LAN) is a group of computers or other devices interconnected within a single, limited area, typically via Ethernet or Wi-Fi.

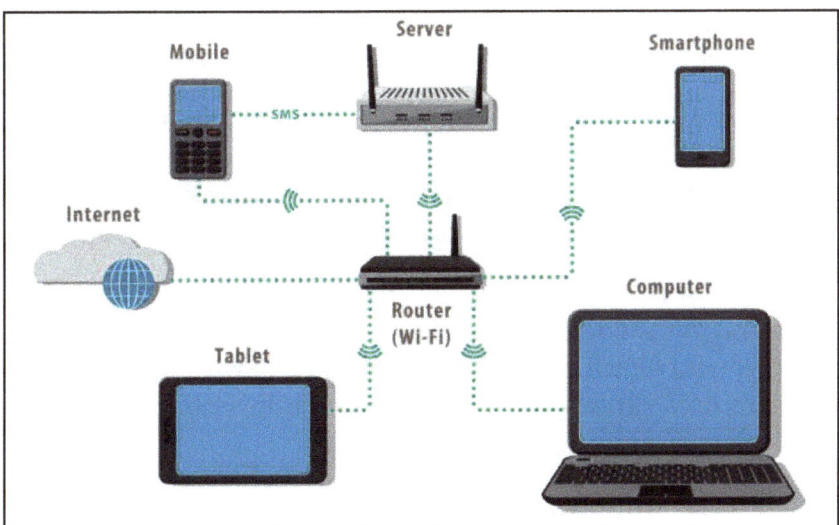

A LAN is a computer network that consists of access points, cables, routers, and switches that enable devices to connect to web servers and internal servers within a single building, campus, or home network, and to other LANs via Wide Area Networks (WAN) or Metropolitan Area Network (MAN). Devices on a LAN, typically personal computers and workstations, can share files and be accessed by each other over a single Internet connection.

A router assigns IP addresses to each device on the network and facilitates a shared Internet connection between all the connected devices. A network switch connects to the router and facilitates communication between connected devices, but does not handle Local Area Network IP configuration or sharing Internet connections. Switches are ideal tools for increasing the number of LAN ports available on the network.

Basic Layouts of Local Area Networks

The Local Area Network layout, also known as Local Area Network topology, describes the physical and logical manner in which devices and network segments are interconnected. LANs are categorized by the physical signal transmission medium or the logical manner in which data travels through the network between devices, independent of the physical connection.

LANs generally consist of cables and switches, which can be connected to a router, cable modem, or ADSL modem for Internet access. LANs can also include such network devices as firewalls, load balancers, and network intrusion detection.

Logical network topology examples include twisted pair Ethernet, which is categorized as a logical bus topology, and token ring, which is categorized as a logical ring topology. Physical network topology examples include star, mesh, tree, ring, point-to-point, circular, hybrid, and bus topology networks, each consisting of different configurations of nodes and links.

How Does Local Area Network Work

The function of Local Area Networks is to link computers together and provide shared access to printers, files, and other services. Local area network architecture is categorized as either peer-to-peer or client-server. On a client-server local area network, multiple client-devices are connected to a central server, in which application access, device access, file storage, and network traffic are managed.

Applications running on the Local Area Network server provide services such as database access, document sharing, email, and printing. Devices on a peer-to-peer local area network share data directly to a switch or router without the use of a central server.

LANs can interconnect with other LANs via leased lines and services, or across the Internet using virtual, private network technologies. This system of connected LANs is classified as a Wide Local Area Network or a metropolitan area network. Local Area and Wide Area Networks differ in their range. An Emulated Local Area Network enables routing and data bridging an Asynchronous Transfer Mode (ATM) network, which facilitates the exchange of Ethernet and token ring network data.

How to Design a Local Area Network

The first step in Local Area Network design is determining network needs. Before building a Local Area Network, identify the number of devices, which determines the number of ports required. A switch can extend the number of ports as the number of devices increases.

In order to connect devices wirelessly, a router is required to broadcast a wireless LAN. A router is also required to establish an internet connection for devices on the network. The distance between hardware devices should be measured in order to determine the length of cables required. Switches can connect cables for very long distances.

The setup simply requires connecting the router to a power source, connecting the modem to the router, connecting the switch to the router (if using), and connecting the devices to the open LAN ports on the router via Ethernet. Next, set up one computer as a Dynamic Host Configuration Protocol server by installing a third-party utility. This will enable all of the connected computers to easily obtain IP addresses. Turn on "Network Discovery" and "File and Printer Sharing" capabilities.

For wireless Local Area Network Installation, start by connecting the computer into one of the router's LAN ports via Ethernet. Enter the router's IP address into any Web Browser and log in with the network administrator account when prompted for a username and password. Open the "Wireless" section in the router settings and change the name of the network in the "SSID" field.

Enable "WPA-2 Personal" as the security or authentication option. Create a password under ""Pre-Shared Key," ensure that the wireless network is "enabled," save changes, restart the router, and connect wireless devices to the wireless network, which should appear on the available network list of devices within range.

Characteristics of wireless Local Area Network include: high capacity load balancing, scalability, network management system, role-based access control, indoor and outdoor coverage options, performance measuring abilities, mobile device management, web content and application filtering, roaming, redundancy, wireless Local Area Network Application prioritization, network switching, and network firewalls.

A common Local Area Network issue is a disabled Local Area Network adapter or adapter error, which can be caused by faulty network adapter settings or by VPN software. Typical solutions include: updating the network adapter driver, resetting the network connection, and checking WLAN AutoConfig dependency services.

How to Secure a Local Area Network

The majority of Local Area Network problems and solutions are concerned with the matter of security. There are a variety of strategies for designing a secure Local Area Network. A common approach is to install a firewall behind a single access point, such as a wireless router. Another valuable measure is to use security protocols such as WPA (Wi-Fi Protected Access) or WPA2 for password encryption on incoming Internet traffic.

Implementing specialized authentication policies enables network administrators to inspect and filter network traffic in order to prevent unauthorized access. Specific access points can be secured with the use of technologies such as VPNs. Internal Local Area Network security can be managed by installing antivirus or anti-malware software.

Virtual Local Area Network

A Virtual Local Area Network (VLAN) is a logical grouping of devices that can assemble together collections of devices on separate physical LANs, and is configured to communicate as if the devices were attached to the same wire. This enables network administrators to easily configure a single switched network to match the security and functional requirements of their systems without requiring any additional cables or significant changes to the current network infrastructure. VLANs are categorized as Protocol VLAN, Static VLAN, or Dynamic VLAN.

Importance of Local Area Network

There are several advantages of Local Area Networks in business:

- Reduced Costs: LANs present a significant reduction in Local Area Network hardware costs and efficient resource pooling.

- Increased Storage Capacity: By pooling all data into a central data storage server, the number of storage servers required is decreased and the efficiency of operations is increased.

- Optimized Flexibility: Data can be accessed by any device from anywhere via Internet connection.

- Streamlined Communication: Files and messages can be transferred in real time and accessed easily from anywhere on any device.

Campus Area Network

Campus Area Network (CAN) is a group of interconnected Local Area Networks (LAN) within a limited geographical area like school campus, university campus, military bases, or organizational campuses and corporate buildings etc. A Campus Area Network is larger than Local Area Network but smaller than Metropolitan Area Network (MAN) and Wide Area Network (WAN).

This Campus Area Network also called as Corporate Area Network. Sometimes this network is also referred as Residential Network or ResNet as it is only used by residents of specific campus only. Campus Area Network is network of interconnected Local Area Networks where these LANs are connected via Switches and routers and create a single network like CAN. Campus Area Network covers areas of around 1 to 5 km range and it can be both wired and wireless connectivity.

Example of CAN: Let's think about a university where university networks interconnect academic building, admission building, library, account section, examination section, placement section etc of an institution when connected with each other combine to form Campus Area Network (CAN). The below figure illustrates a Campus Area Network.

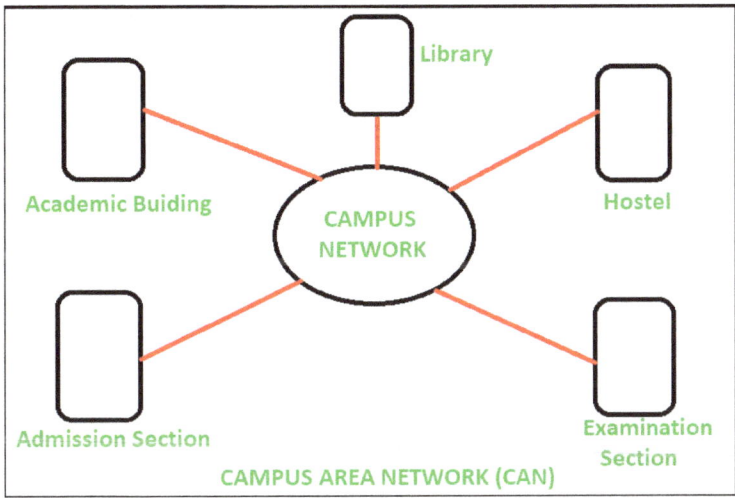

CAMPUS AREA NETWORK (CAN)

Infrastructure of CAN

Within a limited geographical area, LANs are interconnected with help of Switches and Routers and connects buildings to buildings of a single campus where all networking resources like wiring, hubs, switches, routers etc are owned by organization itself. In this, they use same kind of technologies like Local Area Network only interconnection between different buildings is there. Nodes in a campus network are interconnected by means of Optical fiber media, i.e., Fiber optics and takes advantage of 10-Gigabit Ethernet technology. Besides this 10-Gigabit ethernet technology, Wi-Fi hotspots and hot zones are different ways of accessing network.

Advantages of Campus Area Network

There are several benefits of Campus Area Network (CAN), such as:

- Affordability: In CAN network, to use some hardware devices of networking such as hub, routers, switches, cable, bridge etc.

- Easy accessibility of data: Multiple departments of campus are connected to each other in CAN networks. So message is fired one time, and it transferred to all nodes easily.

- Wireless medium: Wireless connections are used to link various offices and buildings with single organization.

- Higher speed: It is capable to transfer huge files with higher speed over entire computer network via internet.

- Protection: CAN Network is combination of multiple LAN networks, and it takes form a single entity. In CAN network, firewall or proxy server are used for security purpose from unauthorized access.

- Share internet connection: Single ISP (Internet Service Provider) is used by different client machines.

Disadvantages of Campus Area Network (CAN)

Campus Area Network has some limitations such as:

- Bound to some connecting nodes: It does not support maximum number of nodes. It can connect up to 64 nodes due to electrical loading.

- Campus Area Network (CAN)'s maintenance is more costly to another network such as LAN, SAN, WAN etc.

- It can support up to 40 meter length.

- In CAN network, to seem undesirable interactions in between all nodes.

Virtual Private Networks

Virtual means not real or in a different state of being. In a VPN, private communication between two or more devices is achieved through a public network the Internet. Therefore, the communication is virtually but not physically there.

- Private: Private means to keep something a secret from the general public. Although those two devices are communicating with each other in a public environment, there is no third party who can interrupt this communication or receive any data that is exchanged between them.

- Network: A network consists of two or more devices that can freely and electronically communicate with each other via cables and wire. A VPN is a network. It can transmit information over long distances effectively and efficiently.

The term VPN has been associated in the past with such remote connectivity services as the (PSTN), Public Switched Telephone Network but VPN networks have finally started to be linked with IP-based data networking. Before IP based networking corporations had expended considerable amounts of time and resources, to set up complex private networks, now commonly called Intranets. These networks were installed using costly leased line services, Frame Relay, and ATM to incorporate remote users. For the smaller sites and mobile workers on the remote end, companies supplemented their networks with remote access servers or ISDN.

Small to medium-sized companies, who could not afford dedicated leased lines, used low-speed switched services. As the Internet became more and more accessible and bandwidth capacities grew, companies began to put their Intranets onto the web and create what are now known as Extranets to link internal and external users. However, as cost-effective and quick-to-deploy as the Internet is, there is one fundamental problem – security. Today's VPN solutions overcome the security factor using special tunneling protocols and complex encryption procedures, data integrity and privacy is achieved, and the new connection produces what seems to be a dedicated point-to point connection. And, because these operations occur over a public network, VPNs can cost significantly less to implement than privately owned or leased services. Although early VPNs required extensive expertise to implement, technology has matured to a level where deployment can be a simple and affordable solution for businesses of all sizes.

Simply put, a VPN, Virtual Private Network, is defined as a network that uses public network paths but maintains the security and protection of private networks. For example, Delta Company has two locations, one in Los Angeles, CA (A) and Las Vegas, Nevada (B). In order for both locations to communicate efficiently, Delta Company has the choice to set up private lines between the two locations. Although private lines would restrict public access and extend the use of their bandwidth, it will cost Delta Company a great deal of money since they would have to purchase the communication lines per mile. The more viable option is to implement a VPN. Delta Company can hook their communication lines with a local ISP in both cities. The ISP would act as a middleman, connecting the two locations. This would create an affordable small area network for Delta Company.

VPNs were broken into 4 categories:

- Trusted VPN: A customer "trusted" the leased circuits of a service provider and used it to communicate without interruption. Although it is "trusted" it is not secured.

- Secure VPN: With security becoming more of an issue for users, encryption and decryption was used on both ends to safeguard the information passed to and fro. This ensured the security needed to satisfy corporations, customers, and providers.

- Hybrid VPN: A mix of a secure and trusted VPN. A customer controls the secure parts of the VPN while the provider, such as an ISP, guarantees the trusted aspect.

- Provider-provisioned VPN: A VPN that is administered by a service provider.

VPN Topology

To begin using a VPN, an Internet connection is needed; the Internet connection can be leased from an ISP and range from a dial up connection for home users to faster connections for businesses. A specially designed router or switch is then connected to each Internet access circuit to provide access from the origin networks to the VPN. The VPN devices create PVCs (Permanent Virtual Circuit- a virtual circuit that resembles a leased line because it can be dedicated to a single user) through tunnels allowing senders to encapsulate their data in IP packets that hide the underlying routing and switching infrastructure of the Internet from both the senders and receivers.

The VPN device at the sending facility takes the outgoing packet or frame and encapsulates it to move through the VPN tunnel across the Internet to the receiving end. The process of moving the packet using VPN is transparent to both the users, Internet Service Providers and the Internet as a whole. When the packet arrives on the receiving end, another device will strip off the VPN frame and deliver the original packet to the destination network.

VPNs operate at either layer 2 or layer 3 of the OSI model (Open Systems Interconnection). Layer-2 VPN uses the layer 2 frame such as the Ethernet while layer-3 uses layer 3 packets such as IP. Layer-3 VPN starts at layer 3, where it discards the incoming layer-2 frame and generates a new layer-2 frame at the destination. Two of the most widely used protocols for creating layer-2 VPNs over the Internet are: layer-2 tunneling protocol (L2TP) and point-to-point tunneling protocol (PPTP). The newly emerged protocol, called Multiprotocol Label Switching (MPLS) is used exclusively in layer-3 VPNs.

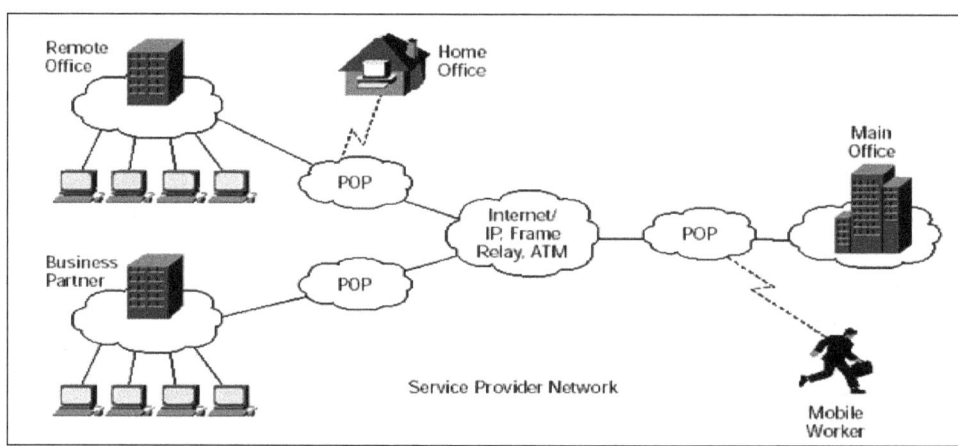

Types of VPNs

There are currently three types of VPN in use: remote access VPN, intranet VPN, extranet VPN. Remote access VPNs enables mobile users to establish a connection to an organization server by using the infrastructure provided by an ISP (Internet Services Provider). Remote access VPN allows users to connect to their corporate intranets or extranets wherever or whenever is needed. Users have access to all the resources on the organization's network as if they are physically located in organization. The user connects to a local ISP that supports VPN using plain old telephone services (POTS), integrated services digital network (ISDN), digital subscriber line (DSL), etc. The VPN device at the ISP accepts the user's login, then establishes the tunnel to the VPN device at the organization's office and finally begins forwarding packets over the Internet. Remote access VPN offers advantages such as:

- Reduced capital costs associated with modem and terminal server equipment.

- Greater scalability and easy to add new users.

- Reduced long-distance telecommunications costs, nationwide toll-free 800 numbers is no longer needed to connect to the organization's modems.

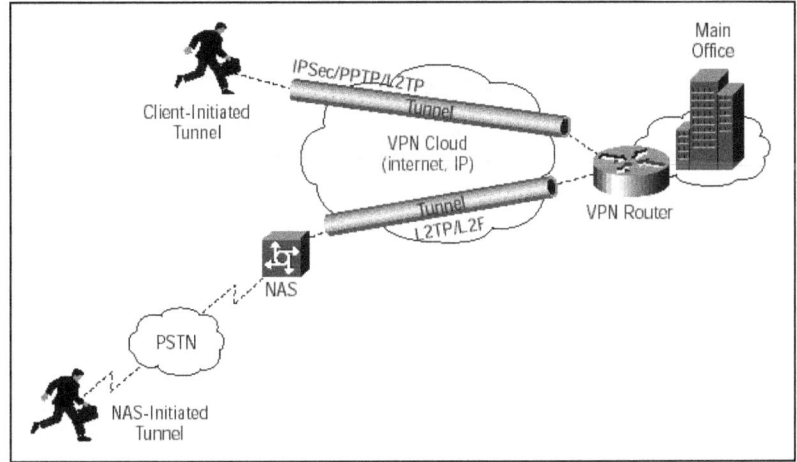

Remote Access VPNs.

Intranet VPNs, provides virtual circuits between organization offices over the Internet. They are built using the Internet, service provider IP, Frame Relay, or ATM networks. An IP WAN infrastructure uses IPSec or GRE to create secure traffic tunnels across the network. Benefits of an intranet VPN include the following:

- Reduced WAN bandwidth costs, efficient use of WAN bandwidth.

- Flexible topologies.

- Congestion avoidance with the use of bandwidth management traffic shaping.

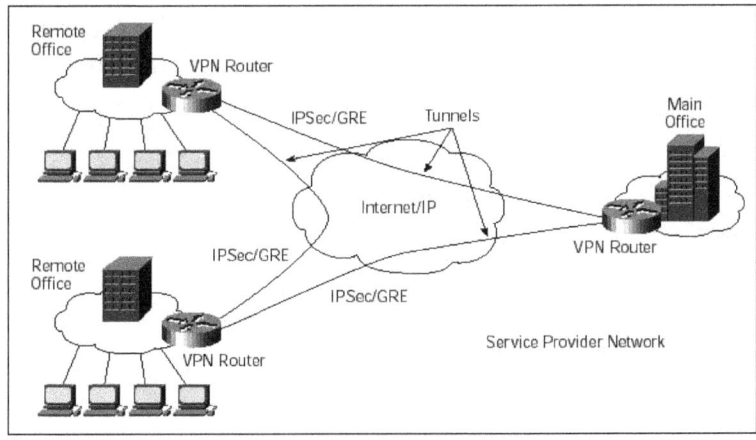

Intranet VPNs.

The concept of setting up extranet VPNs is the same as intranet VPN. The only difference is the users. Extranet VPN are built for users such as customers, suppliers, or different organizations over the Internet.

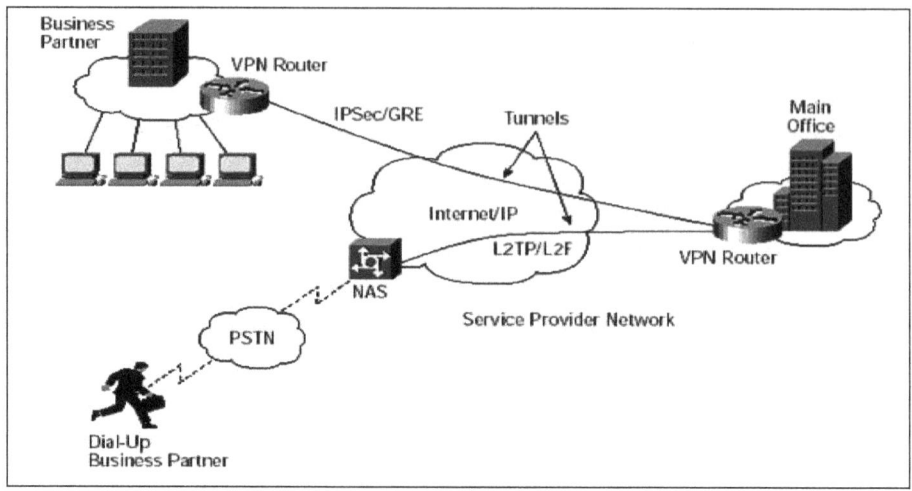

Extranet VPNs.

Components of the VPN

In order for a VPN to be beneficial a VPN platform needs to be reliable, manageable across the enterprise and secure from intrusion. The VPN solution also needs to have Platform Scalability – the ability to adapt the VPN to meet increasing requirements ranging from small office configuration to large enterprise implementations. A key decision the enterprise should make before starting their implementation is to consider how the VPN will grow to meet the requirement of the enterprise network and if VPN will be compatible with the legacy networks already in place.

Security

Companies need to keep their VPNs secure from tampering and unauthorized users. Some examples of technologies that VPN's use are; IP Security (IPSec), Point-to-Point Tunneling Protocol (PPTP), Layer 2 Tunneling Protocol and Multiprotocol Label Switching (MPLS) along with Data Encryption Standard (DES), and others to manage security.

PPTP uses Point-to-Point Protocol (PPP) to provide remote access that can be tunneled through the Internet to a desired site. Tunneling allows senders to encapsulate their data in IP packets that hide the routing and switching infrastructure of the Internet from both senders and receivers to ensure data security against unwanted viewers, or hackers. PPTP can also handle Internet packet exchange (IPX) and network basic input/output system extended user interface (NetBEUI).

PPTP is designed to run on the Network layer of the Open systems interconnection (OSI). It uses a voluntary tunneling method, where connection is only established when the individual user request to logon to the server. PPTP tunnels are transparent to the service provider and there is no advance configuration required by the Network Access Server, this allows PPTP to use multiple service providers without any explicit configuration. For example, the client dials up to the ISP and makes a PPP session. Then, the client dials again to the same PPP session, to contact with the destination remote access server (RAS). After contact is made with the RAS, packets are then tunneled through the new connection and the client is now connected to the corporate server virtually.

Layer Two Tunneling Protocol (L2TP) exists at the data link layer of the OSI model. L2TP is a combination of the PPTP and Layer two Forwarding (L2F). (Layer two forwarding was also designed for traffic tunneling from mobile users to their corporate server. L2F is able to work with media such as frame relay or asynchronous transfer mode (ATM) because it does not dependent on IP. L2F also uses PPP authentication methods for dial up users, and it also allows a tunnel to support more than one connection.) L2TP uses a compulsory tunneling method, where a tunnel is created without any action from the user, and without allowing the user to choose a tunnel. A L2TP tunnel is dynamically established to a predetermined end-point based on the Network Access Server (NAS) negotiation with a policy server and the configured profile. L2TP also uses IPSec for computer-level encryption and data authentication.

IPSec uses data encryption standard (DES) and other algorithms for encrypting data, public-key cryptography to guarantee the identities of the two parties to avoid man-in-the-middle attack, and digital certificates for validating public keys. IPSec is focused on Web applications, but it can be used with a variety of application-layer protocols. It sits between IP at the network layer and TCP/UDP at the transport layer. Both parties negotiated the encryption technique and the key before data is transferred. IPSec can operate in either transport mode or tunnel mode.

- In tunnel model, intruders can only see where the end points of the tunnel are, but not the destinations of the packet and the sources. IPSec encrypts the whole packet and adds a new IP packet that contains the encrypted packet. The new IP packet only identifies the destination's encryption agent. When the IPSec packet arrives at the encryption agent, the new encrypted packet is stripped and the original packet continues to its destination.

- In Transport mode IPSec leaves the IP packet header unchanged and only encrypts the IP payload to ease the transmission through the Internet. IPSec here adds an encapsulating security payload at the start of the IP packet for security through the Internet. The payload header provides the source and destination addresses and control information.

Multiprotocol Label Switching (MPLS) uses a label swapping forwarding structure. It is a hybrid architecture which attempts to combine the use of network layer routing structures and per-packet switching, and link-layer circuits and per-flow switching. MPLS operates by making the inter-switch transport infrastructure visible to routing and it can also be operated as a peer VPN model for switching a variety of link-layer and layer 2 switching environments. When the packets enter the MPLS, it is assigned a local label and an outbound interface based on the local forwarding decision. The forwarding decision is based on the incoming label, where it determines the next interface and next hop label. The MPLS uses a look up table to create end-to-end transmission pathway through the network for each packet.

Packet authentication prevents data from being viewed, intercepted, or modified by unauthorized users. Packet authentication applies header to the IP packet to ensure its integrity. When the receiving end gets the packet, it needs to check for the header for matching packet and to see if the packet has any error.

User authentication is used to determine authorized users and unauthorized users. It is necessary to verify the identity of users that are trying to access resources from the enterprise network before they are given the access. User authentication also determines the access levels; data retrieved or viewed by the users, and grant permission to certain areas of the resources from the enterprise.

Appliances: Intrusion Detection Firewalls

Firewalls monitors traffic crossing network parameter, and protect enterprises from unauthorized access. The organization should design a network that has a firewall in place on every network connection between the organization and the Internet. Two commonly used types of firewalls are packet-level firewalls and application-level firewalls.

Packet-level firewall checks the source and destination address of every packet that is trying to passes through the network. Packet-level firewall only lets the user in and out of the organization's network only if the users have an acceptable packet with the correspondent source and destination address. The packet is checked individually through their TCP port ID and IP address, so that it knows where the packet is heading. Disadvantage of packet-level firewall is that it does not check the packet contents, or why they are being transmitted, and resources that are not disabled are available to all users.

Application-level firewall acts as a host computer between the organization's network and the Internet. Users who want to access the organization's network must first log in to the application-level firewall and only allow the information they are authorized for. Advantages for using application-level firewall are: users access level control, and resources authorization level. Only resources that are authorized are accessible. In contrast, the user will have to remember extra set of passwords when they try to login through the Internet.

Management

Managing security policies, access allowances, and traffic management. VPN's need to be flexible to a companies management, some companies chooses to manage all deployment and daily operation of their VPN, while others might choose to outsource it to service providers.

Productivity and Cost Benefit

In terms of productivity VPN's have come a long way. In the past, concerns over security and manageability overshadowed the benefits of mobility. Smaller organizations had to consider the additional time and cost associated with providing IT support to employees on the move. Larger companies worried, with good cause, about the possibility that providing mobile workers with remote network access would inadvertently provide hackers with a "back door" entry to corporate information resources. But as end-user technologies like personal digital assistants (PDAs) and cell phones have made mobility more compelling for employees, technology advances on the networking side have helped address IT concerns. With these advancements in technology comes better productivity. VPN's have become increasingly important because they enable companies to create economical, temporary, secure communications channels across the public Internet so that mobile workers can connect to the corporate LAN. VPN's Benefit a company in the following ways:

- Extends Geographic Connectivity: A VPN connects remote workers to central resources, making it easier to set up global operations.

- Boosts Employee Productivity: A VPN solution enables telecommuters to boost their productivity by 22%-45% by eliminating time-consuming commutes and by creating uninterrupted time for focused work.

- Improves Internet Security: An always-on broadband connection to the Internet makes a network vulnerable to hacker attacks. Many VPN solutions include additional security measures, such as firewalls and anti-virus checks to counteract the different types of network security threats.

- Scales Easily: A VPN allows companies to utilize the remote access infrastructure within ISPs. Therefore, companies are able to add a virtually unlimited amount of capacity without adding significant infrastructure.

Even though VPN's are a cheaper way of having remote users connect to a company's network over the Internet there are still costs associated with implementing the VPN. Some of the typical costs include hardware, ISP subscription fees, network upgrading costs and end user support costs. These costs aren't standard they vary depending on many factors, some of which include, size or corporation, number of remote users, type of network systems already in place and Internet Service Provider source. When it comes to decision making time IT managers or Executive officers should take these costs into consideration. Also these decision makers must decide whether to develop their VPN solution in house or to outsource to a total service provider. There are a few ways to approach this topic:

- In House Implementation: Companies decide that for their needs an in-house solution is all they need. These companies would rather set up individual tunnels and devices one at a time and once this is established the company can have their own IT staff take care of the monitoring and upkeep.

- Outsourced Implementation: Companies can choose to outsource if they are large scaled or lack the IT staff to fully implement an in house VPN. When a company outsources the service provider usually designs the VPN and manages it on the company's behalf.

- Middle Ground Implementation: Some companies would rather have a service provider install the VPN but has their IT staff monitored the specifics such as tunnel traffic. This type of implementation is a compromise between a company and the service provider.

After Implementation the company must make sure that it has adequate support for its end users. That's where quality of service comes in.

Quality of Service (QOS)

Users of a widely scattered VPN do not usually care about the network topology or the high level of security/encryption or firewalls that handle their traffic. They don't care if the network implementers have incorporated IPSec tunnels or GRE tunnels. What they care about is something more fundamental, such as: "Do I get acceptable response times when I access my mission critical applications from a remote office?"

Acceptance levels for delays vary. While a user would be willing to put up with a few additional seconds for a file transfer to complete, the same user would have less tolerance for similar delays when accessing a database or when running voice over an IP data network.

QoS (Quality of Service) aims to ensure that your mission critical traffic has acceptable performance. In the real world where bandwidth is limited and diverse applications from videoconferencing to ERP database lookups must all strive for scarce resources, QoS becomes a vital tool to ensure that all applications can coexist and function at acceptable levels of performance.

Quality of Service (QOS) is a key component of any VPN service. In MPLS/BGP VPNs, existing L3 QoS capabilities can be applied to labeled packets through the use of the "experimental" bits in the header, or, where ATM is used as the backbone, through the use of ATM QoS capabilities. Traffic engineering could even be used to establish LSPs with particular QoS characteristics between particular pairs of sites, if that is desirable. Where an MPLS/BGP VPN spans multiple SPs, the architecture described may be useful. An SP may apply either intserv or diffserv capabilities to a particular VPN, as appropriate.

The Future of VPN

As more and more businesses demand a higher level of network access, the business is migrating from a private network environment to a new model in which information is distributed throughout the enterprise network. Thus, expanding their network in the near future and actually seeing the benefits of using the Internet as the backbone to create Virtual Private Networks (VPN). VPN is designed to meet the demands for information access in a secure, cost-effective environment.

Multi-vendor interoperability for VPN is crucial in today's networking environment due to the nature of business successes, the need to extend corporate networks to contractors and partners, and the diverse equipment within company networks. The Microsoft Windows operating system has integrated VPN technology that helps provide secure, low-cost remote access and branch office connectivity over the internet.

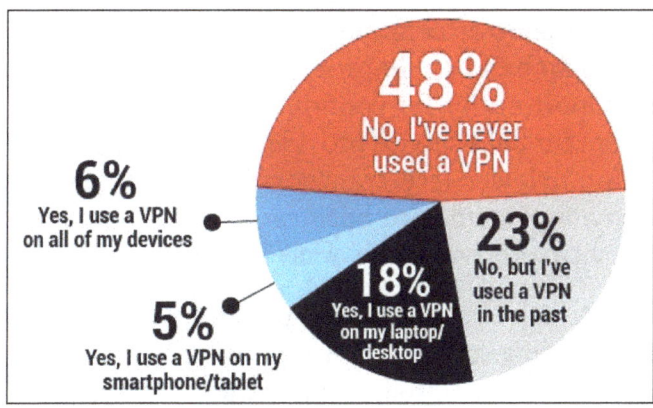

Companies with VPN.

The future is in integrated VPNs which depend on how VPNs industry will improve their unique qualities that will enable consumers to communicate effectively with other consumers. Therefore, a VPN creates a large, multi-site, company-wide data network which allows for every device to be uniquely addressed from anywhere on the network. This means that central resources can be accessed from any site in the organization or from any Internet-connected location around the world. The technical problems involved in connecting hundreds of remote sites to a central network are extensive. It often involves the purchase of very expensive high-density backbone routers or the use of costly frame-relay services. These systems are seldom easy to support and often require specialist skills. Also, it depends on the ability of intranets and extranets to deliver on their promises. First of all VPN companies must consider to cost saving for servicing of VPNs. Generally speaking the more the companies supply cheaper cost of services, the more products or demands increase for them on the markets. Therefore, they will earn high profit then spend a lot of money for developing much higher quality VPN. Here is a diagram for U.S. companies with IP VPN.

Demand for VPN has been increasing even though economy is going down and especially IT business companies have not succeeded at present. More then 20 percents of companies will plan to have IP VPN services in the future so those in near future more than 70 percents of companies are going to use IP VPN services. More companies will adopt IP VPN services and increasing more demand in the U.S. Also many companies have been using IP VPN for remote access as LAN.

The companies for servicing VPN will consider meeting consumer's demands that is voice over IP and other VPN as VOIP VPN. Currently very a few companies have been using this VPN and a few companies will plan to use it in the future. However, contrary to their demands, most produces are standing on difficult situation for improving VOIP VPN because the voice is a kind of special requirement of low latency and jitter. Most of people will continue to use voice communication by telephone that is successfully improving with low costs.

The 21st century invites new ways of viewing the communication networks. Companies that previously managed their own communications requirements are uniting with service providers that can help build up, improve, and manage their networks on a global scale. This opens up opportunities for continued growth, increased profitability, and the greatest achievement for both service providers and subscribers. In the past, service providers drew attention to lower-level transport, such as leased lines and frame relay. Nowadays, service providers team with business customers to meet their networking requirements through virtual private networks (VPNs).

VPNs are the source of future services. When properly implemented, they can simplify network operations while reducing capital expenses. For most companies, the starting point is to connect widely separated workgroups in an efficient, moneymaking manner. From there, service providers can influence the main technology as a foundation for offering additional services such as application hosting, videoconferencing, and packet telephony.

VPN help service providers build customer loyalties while delivering network services that are valuable to their customers' business operations. This indicates an opportunity to capture new customers, as companies switch from yesterday's data communications strategies to today's more comprehensive at hand solutions.

Client/Server Networks

Client server network is such model where one side server machine delivers the various services to other side client machine for grabbing those services. So this type of model is known as the "Client-Server Networking Model".

In this network, entire network is controlled by centralized powerful computer; it is called the "Server". When client sends the requests for grabbing many services, then server terminal is getting to open the window for incoming all requests. Server is also capable to perform all types of massive operations such as security and network management.

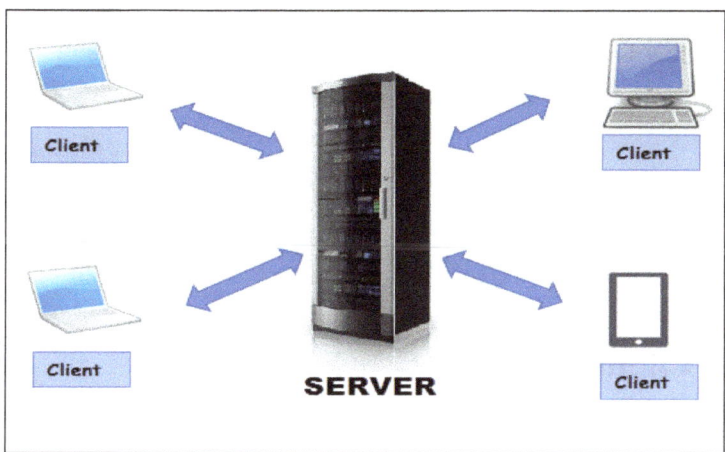

Diagram of Client-Server Network.

Server has right to handle all types of network resources like as files, folders, directories, various applications, and other shared terminals such as printer etc. If, anytime any client wants to need those services, then it firstly takes all permissions from server side through sending request.

This type of Server is capable to deliver several services for various client machines not for specific one client. So client server network is enabled with many-to-one relationship model. Mostly Local area networks are designed on the base of client server model relationship. Client Server network was getting more popularity in late 1980, but in 1990 various applications were swapped from centralized minicomputer and mainframe system to computer network system of your personal computers.

How to Work Client Server Model

- Client: Client is a service requester that sends the requests to server for obtaining of necessary services.

- Server: Server is a high performer computer machine that delivers many functionality for another device or program.

Client (user) feeds the URL (Uniform Resource Locator) of any website, then browser sends the requests to DNS (DOMAIN NAME SYSTEM) server. DNS server machine find out the appropriate address of your web server. DNS server is responsible to responds with using of IP (Internet Protocol) address of your web server. Then, Browser fires HTTP/HTTPS requests to IP of web server. In the next step, all needed files of your website are sent by server. Finally, browser is getting to render all files and your website is visible on your system, and its rendering activities are completed with using of DOM, CSS and LS engine.

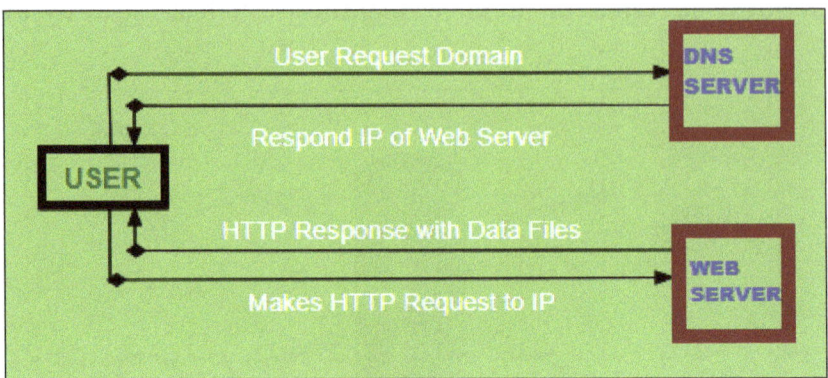

Advantages of Client Server Network

There are few benefits of client server architecture, such as:

- Client server network has fully right to control all activities of entire network centralized.

- All data is saved on the central area.

- All network devices can be handled centrally.

- All concern like as Backups and network protection also can be managed centrally.

- All users also have right to access entire shard files which are stored centrally.

- Users are able to access all data anytime and anywhere, so you have not any place boundation.

- It can be scaled that means as per the requirement its size can be expanded.

- It provides the Integration of services that means It grants permission of all your client to access corporate data with using of own terminal, and to eliminate all unnecessary log in permissions.

- It allows users to share all resources on the other different platforms and locations.

- Client server network is designed on the base of distributed model that means to replace, repair, any updating and relocate server without affecting of client.

- It is capable to bear massive usage.

- Client server network has best management to keep all records of entire files, so all users can find any file easily.

- It allows to all users to decrease the data replication for their applications.

Disadvantages of Client Server Network

- If, main server gets halt then entire system will be failed.

- Client server network is need special network operating system.

- More expensive to configure their hardware and software components.

- To need well qualified technical staff for maintaining the server.

- Traffic Congestion Problem: If large number of client try to send requests at the same time frame then "Traffic Congestion Problem" can be created.

- Its cost is not affordable for normal users.

Four Types of Client Server Network

- Web Servers: Web server likes as high performance computer system that can host multiples websites. On this server, to install different types of web server software like as Apache or Microsoft IIS, which delivers access to hosted several websites on the internet, and these servers are linked with internet through higher speed connection that delivers ultra data transmission rates.

- Mail Servers: Email servers help to send and receive all emails. Some software are run on the mail server which allow to administrator to create and handle all email accounts for any domain that is hosted on the server. Mail servers use the some protocols for sending and receiving emails such as SMTP, IMAP, and POP3. SMTP protocol helps to fire messages and manages all outgoing email requests. IMAP and POP3 help to receive all messages and handle all incoming mails.

- File Servers: File server is dedicated systems that allow users to access for all files. It works like as centralized file storage location, and it can be accessed by several terminal systems.

- DNS: DNS stands for "Domain Name Server", and it has huge database of different types of public IP addresses, and they link with their hostnames.

These types of server help to deliver all resources (like as files, directories, shared devices such as applications and printers) to client terminal like as PCs, smart phones, PDAs, laptops, tablets etc.

Peer-to-Peer Networks

A peer-to-peer network is a technology that allows you to connect two or more computers to one system. This connection allows you to easily share data without having to use a separate server for your file-sharing. Each end-computer that connects to this network becomes a 'peer' and is allowed to receive or send files to other computers in its network. This enables you to work collaboratively to perform certain tasks that need group attention, and it also allows you to provide services to another peer.

Importance of P2P Network

This technology is a big advancement in the IT world since it enhances companies' efficiencies. Peer-to-peer networking has also made communication far easier than it used to be, which is why many businesses have quickly adapted to this technology.

Another primary benefit of this software is that it allows companies to stay connected within the workplace, increasing a business's overall efficiency. Additionally, once your entire computer system has synced, you may be able to keep your data collected and checked in one place, advancing the overall security of your data.

Operating P2P Network

This software works to prohibit the sharing of information until a peer grants you access. So, if a computer wants to access files from a different computer, it will need permission to access it. A peer is in full control of their files and has the option to allow or deny access to another peer. For instance, if user X wants to see the files of user Y, user X must first seek permission from user Y to reach the files. User Y can deny this access or accept it and provide the password to user X. Other key uses of a P2P network include:

- File sharing: The use of P2P in file sharing is extremely convenient for businesses. P2P networking can also save you money with this feature because it eliminates the need to use another intermediate server to transfer your file.

- Direct messaging: Another great feature of P2P networks is that they provide a secure and quick way to communicate. This is possible when you follow a certain protocol that permits encryption at both peer computers and gives easy messaging tools.

- Collaboration: The easy sharing of files also helps build stronger collaboration with the rest of your colleagues. You can perform an extensive amount of work with this tool and do certain tasks that need thorough checking.

- System back-up: Another strong P2P networking feature is that even if one computer shuts down for some reason, the rest of the computers continue working.

This networking only stops when all the peer computers shut down at once. So, the chances of losing data are minimal and the chances of securing file sharing are maximum. This feature also increases the reliability of P2P networking.

How to keep your P2P Network Secure

Keeping your P2P networking systems secure increases your company's protection and allows it to operate more effectively. Follow these steps to help you properly secure your P2P networking systems:

- Share and download legal files: When using the P2P network, try to download only legal files. You can use the P2P network to download music, video and work-related files to share with other employees. Double-check that the files you're sharing and downloading are all legal files.

- Keep your security up-to-date: Similar to a majority of computer and software technologies, a P2P network is often prone to damage. Use antivirus software to keep your files protected from any damaging viruses.

- Scan all of your downloads: A quick way to keep your P2P network safe is to constantly check and scan your files for viruses before downloading them. This helps ensure your computer is safe from potential harm. If it detects any viruses, immediately report it to the IT staff members at the company.

- Shut down the P2P networking system after use: Try to make sure you correctly shut down the software to avoid unnecessary access of a third person into your files. An unauthorized third person gaining access to your files can be a major security breach in your network. Even if you close the windows after sharing files, someone else can still gain access to your files if the software remains active and isn't shut down.

Usage of Peer-to-Peer Network

Many systems have used P2P networking to make money and keep their systems running smoothly. Because the P2P network is user-friendly, it makes it easier for users to regularly store their data. You can make P2P a part of your business to streamline your data filing and sharing systems. Your business can also earn an income from P2P networking:

- Cloud servers: This is a very common form of using a P2P network because many users share data on their cloud. It helps them in accessing their files from many other sources. The storage facility is another factor for uploading files on a cloud, because even though many applications charge money for additional storage, the current amount of storage offered on the cloud is enough for most users.

- Business communication and messaging applications: Most workplace messaging systems and business collaboration hubs develop a direct way of communicating with colleagues and coworkers. Similar to how P2P networking systems encrypt texts and calls, many direct chat and video messaging systems use the same method to share information and keep user communication secure from any third parties.

- Content distribution: Many applications use the encryption provided by the P2P network to devise a way to share files on a larger scale. Most content distribution applications work based on the P2P network to prevent a file and data from piracy. The encryption of these messages has proven to increase the users' overall safety and protection. This way the users can sell and efficiently use these sites to convey messages.

  Many people enjoy content distribution software because they can access files and videos without having to wait a significant amount of time. Additionally, the increased traffic makes these sites more cost-effective, since there's no need for a central server.

Advantages of Peer-to-Peer Network

- Cost: The overall cost of building and maintaining a peer to peer network is relatively inexpensive. The setup cost has been greatly reduced due to the fact that there is no central configuration. Moreover for the windows server, there is no payment required for each of the users on the network. The payment should be done only once.

- Reliability: Peer to Peer network is not dependent on a centralized system. Which means that the connected computers can function independently with each other? Even if one part of the network fails, it will not disrupt other parts. Only the user will not be able to access those files.

- Implementation: It is generally easy to setup a peer to peer network requiring no advanced knowledge. Only a hub or a switch is needed for the connection. And also since all the connected computers can manage themselves, there should be no many configurations. However it needs some specialized software.

- Scalability: P2P networking has one of the best scalability features. Even if there are extra clients added, the performance of the network will remain the same. Sometimes more users tend to share a single file. For this case, the network will increase the availability of bandwidth.

- Administration: There is no need for any specialized network administrator since all the users are given the right to manage their own system. They can choose what type of files they are willing to share.

- Server Requirement: In peer to peer networking, each connected computers acts as a server and a workstation. Therefore, there is no need to use a dedicated server. All the authorized users can use their respective client computer to access the required files. This can lead to saving more overhead costs.

- Resource Sharing: In P2P networking, the resources are shared equally among all the users. The connected devices can provide and consume resources at the same time. And also this peer to peer networking can be used for locating and downloading online files easily.

Disadvantages of Peer-to-Peer Networking

- Decentralization: Peer to Peer networking lacks the feature of centralization. There is no central server, thus files are stored on individual machines. The entire network accessibility is not in the hands of a single person. This makes it more challenging for the users to locate and find files. If the search is done through each database, the users could waste a lot of time.

- Performance: Performance is another issue faced by a peer to peer network. Once the number of devices connecting the network increases, there will be a performance degrade since each computer is being accessed by other users. Hence, P2P network doesn't work well with growing networks.

- Security: Security for individual files is comparatively less in peer to peer networking. There is no security other than assigning permissions. Even if the permissions are assigned, any person with the access to it will be able to log on. Some users don't even require logging on from their respective workstation.

- Remote Access: In some cases, there can be unsecured types of codes present on a particular terminal. If this is the case, there are possibilities where files on a network will be accessed by remote users without proper permissions. This can lead to a compromised network.

- Backup Recovery: Backup is made way difficult in P2P networks, since the data is not centralized. It is saved on various systems. Therefore, backup needs to be done separately on each computer. Or else there should be a backup system for every computer.

- Virus Attacks: Peer to peer networks is more prone to malware and virus attacks since each connected computers are independent to each other. If one of the computers tends to get virus infected, it could easily spread to the remaining computers even if they are protected through an antivirus or firewall software. Therefore, it is the responsibility of each user to make sure that their system is protected against viruses.

- Illegal Content: Most often peer to peer networks are used to transfer copyrighted contents like movies and music by implementing into torrents. Due to this there is a possibility of internet ban, notice from content writers or even arrest. That is the reason why P2P networks are less preferred among some companies and service providers.

References

- What-is-a-wan-wide-area-network-definition-and-examples-3248989: networkworld.com, Retrieved 29, Jan 2020

- What-is-a-wan-wide-area-network: cisco.com, Retrieved 17, June 2020

- Types-wan-technologies-data-networking: bsimplify.com, Retrieved 13, August 2020

- Networking-fundamentals-metropolitan-area-network-man: quickstart.com, Retrieved 23, April 2020

- Metropolitan-area-network-definition: router-switch.com, Retrieved 25, March 2020

- What-is-campus-area-network-can-definition-advantages-disadvantages: digitalthinkerhelp.com, Retrieved 14, Feb 2020

- What-is-client-server-network-example-advantages-disadvantages: digitalthinkerhelp.com, Retrieved 03, May 2020

3

Modern Wireless Networks

Wireless technology is the technology that allows communication to take place without the need for connecting wires or cables. It is responsible for the creation and widespread adoption of wireless networks. Wireless LAN, wireless access point, wireless WAN, wireless mesh network, etc. are all examples of modern wireless networks. This chapter is a comprehensive summary of modern wireless networks, its affiliated concepts and applications.

Wireless

Wireless communication is the transfer of information or power between two or more points that are not connected by an electrical conductor.

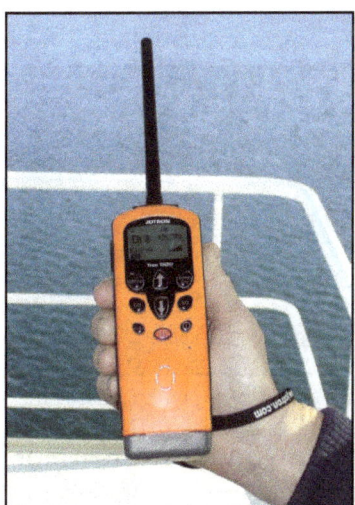

A handheld On-board communication station of the maritime mobile service.

The most common wireless technologies use radio. With radio waves distances can be short, such as a few meters for television or as far as thousands or even millions of kilometers for deep-space radio communications. It encompasses various types of fixed, mobile, and portable applications, including two-way radios, cellular telephones, personal digital assistants (PDAs), and wireless networking. Other examples of applications of radio *wireless technology* include GPS units, garage door openers, wireless computer mice, keyboards and headsets, headphones, radio receivers, satellite television, broadcast television and cordless telephones.

Somewhat less common methods of achieving wireless communications include the use of other electromagnetic wireless technologies, such as light, magnetic, or electric fields or the use of sound.

The term *wireless* has been used twice in communications history, with slightly different meaning. It was initially used from about 1890 for the first radio transmitting and receiving technology, as in *wireless telegraphy*, until the new word *radio* replaced it around 1920. The term was revived in the 1980s and 1990s mainly to distinguish digital devices that communicate without wires, such as the examples listed in the previous paragraph, from those that require wires. This is its primary usage today.

LTE, LTE-Advanced, Wi-Fi, Bluetooth are some of the most common modern wireless technologies.

Introduction

Wireless operations permit services, such as long-range communications, that are impossible or impractical to implement with the use of wires. The term is commonly used in the telecommunications industry to refer to telecommunications systems (e.g. radio transmitters and receivers, remote controls, etc.) which use some form of energy (e.g. radio waves, acoustic energy, etc.) to transfer information without the use of wires. Information is transferred in this manner over both short and long distances.

History

Photophone

Bell and Tainter's photophone, of 1880.

The world's first wireless telephone conversation occurred in 1880, when Alexander Graham Bell and Charles Sumner Tainter invented and patented the photophone, a telephone that conducted audio conversations wirelessly over modulated light beams (which are narrow projections of electromagnetic waves). In that distant era, when utilities did not yet exist to provide electricity and lasers had not even been imagined in science fiction, there were no practical applications for their invention, which was highly limited by the availability of both sunlight and good weather. Similar to free-space optical communication, the photophone also required a clear line of sight between its transmitter and its receiver. It would be several decades before the photophone's principles found their first practical applications in military communications and later in fiber-optic communications.

Early Wireless Work

David E. Hughes transmitted radio signals over a few hundred yards using a clockwork keyed transmitter in 1878. As this was before Maxwell's work was understood, Hughes' contemporaries dismissed his achievement as mere "Induction." In 1885, Thomas Edison used a vibrator magnet for induction transmission. In 1888, Edison deployed a system of signaling on the Lehigh Valley Railroad. In 1891, Edison obtained the wireless patent for this method using inductance (U.S. Patent 465,971).

In 1888, Heinrich Hertz demonstrated the existence of electromagnetic waves, the underlying basis of most wireless technology. The theory of electromagnetic waves was predicted from the research of James Clerk Maxwell and Michael Faraday. Hertz demonstrated that electromagnetic waves traveled through space in straight lines, could be transmitted, and could be received by an experimental apparatus. Hertz did not follow up on the experiments. Jagadish Chandra Bose around this time developed an early wireless detection device and helped increase the knowledge of millimeter-length electromagnetic waves. Later inventors implemented practical applications of wireless radio communication and radio remote control technology.

Radio

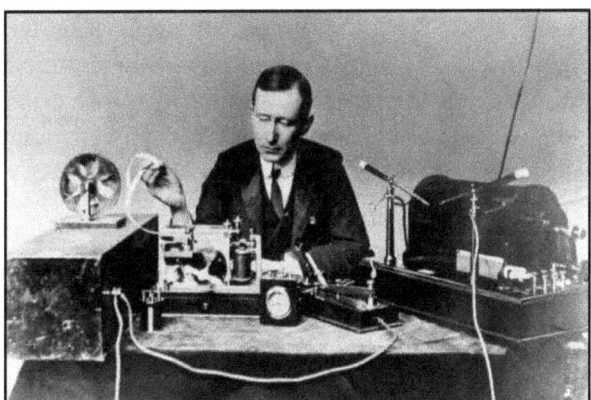

Marconi transmitting the first radio signal across the Atlantic.

The term "wireless" came into public use to refer to a radio receiver or transceiver (a dual purpose receiver and transmitter device), establishing its use in the field of wireless telegraphy early on; now the term is used to describe modern wireless connections such as in cellular networks and wireless broadband Internet. It is also used in a general sense to refer to any operation that is implemented without the use of wires, such as "wireless remote control" or "wireless energy transfer", regardless of the specific technology (e.g. radio, infrared, ultrasonic) used. Guglielmo Marconi and Karl Ferdinand Braun were awarded the 1909 Nobel Prize for Physics for their contribution to wireless telegraphy.

Modes

Wireless communications can be via:

Radio

radio communication, microwave communication, for example long-range line-of-sight via highly directional antennas, or short-range communication.

Free-space Optical

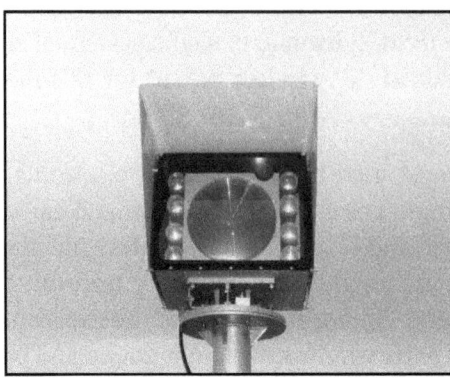

An 8-beam free space optics laser link, rated for 1 Gbit/s at a distance of approximately 2 km. The receptor is the large disc in the middle, the transmitters the smaller ones. To the top and right corner a monocular for assisting the alignment of the two heads.

Free-space optical communication (FSO) is an optical communication technology that uses light propagating in free space to transmit wirelessly data for telecommunications or computer networking. "Free space" means the light beams travel through the open air or outer space. This contrasts with other communication technologies that use light beams traveling through transmission lines such as optical fiber or dielectric "light pipes".

The technology is useful where physical connections are impractical due to high costs or other considerations. For example, free space optical links are used in cities between office buildings which are not wired for networking, where the cost of running cable through the building and under the street would be prohibitive.

Another widely used example is consumer IR devices such as remote controls and IrDA (Infrared Data Association) networking, which is used as an alternative to WiFi networking to allow laptops, PDAs, printers, and digital cameras to exchange data.

Sonic

Sonic, especially ultrasonic short range communication involves the transmission and reception of sound.

Electromagnetic Induction

Electromagnetic induction short range communication and power. This has been used in biomedical situations such as pacemakers, as well as for short-range Rfid tags.

Wireless Services

Common examples of wireless equipment include:

- Infrared and ultrasonic remote control devices.

- Professional LMR (Land Mobile Radio) and SMR (Specialized Mobile Radio) typically used by business, industrial and Public Safety entities.

- Consumer Two-way radio including FRS Family Radio Service, GMRS (General Mobile Radio Service) and Citizens band ("CB") radios.

- The Amateur Radio Service (Ham radio).

- Consumer and professional Marine VHF radios.

- Airband and radio navigation equipment used by aviators and air traffic control.

- Cellular telephones and pagers: provide connectivity for portable and mobile applications, both personal and business.

- Global Positioning System (GPS): allows drivers of cars and trucks, captains of boats and ships, and pilots of aircraft to ascertain their location anywhere on earth.

- Cordless computer peripherals: the cordless mouse is a common example; wireless headphones, keyboards, and printers can also be linked to a computer via wireless using technology such as Wireless USB or Bluetooth.

- Cordless telephone sets: these are limited-range devices.

- Satellite television: Is broadcast from satellites in geostationary orbit. Typical services use direct broadcast satellite to provide multiple television channels to viewers.

Computers

- WiFi.

- Cordless computer peripherals:

    - mouse,

    - headphones,

    - keyboards,

    - printers,

    - USB and,

    - Bluetooth.

- Wireless networking:

    - To span a distance beyond the capabilities of typical cabling,

    - To provide a backup communications link in case of normal network failure,

    - To link portable or temporary workstations,

    - To overcome situations where normal cabling is difficult or financially impractical, or

    - To remotely connect mobile users or networks.

Developers need to consider some parameters involving Wireless RF technology for better developing wireless networks:

- Sub-GHz versus 2.4 GHz frequency trends.

- Operating range and battery life.

- Sensitivity and data rate.

- Network topology and node intelligence.

Applications may involve point-to-point communication, point-to-multipoint communication, broadcasting, cellular networks and other wireless networks, Wi-Fi technology.

Cordless

The term "wireless" should not be confused with the term "cordless", which is generally used to refer to powered electrical or electronic devices that are able to operate from a portable power source (e.g., a battery pack) without any cable or cord to limit the mobility of the cordless device through a connection to the mains power supply.

Some cordless devices, such as cordless telephones, are also wireless in the sense that information is transferred from the cordless telephone to the phone's base unit via some wireless communications link. This has caused some disparity in the usage of the term "cordless", for example in Digital Enhanced Cordless Telecommunications.

Electromagnetic Spectrum

Light, colors, AM and FM radio, and electronic devices make use of the electromagnetic spectrum. The frequencies of the radio spectrum that are available for use for communication are treated as a public resource and are regulated by national organizations such as the Federal Communications Commission in the USA, or Ofcom in the United Kingdom. This determines which frequency ranges can be used for what purpose and by whom. In the absence of such control or alternative arrangements such as a privatized electromagnetic spectrum, chaos might result if, for example, airlines did not have specific frequencies to work under and an amateur radio operator were interfering with the pilot's ability to land an aircraft. Wireless communication spans the spectrum from 9 kHz to 300 GHz.

Applications of Wireless Technology

Mobile Telephones

One of the best-known examples of wireless technology is the mobile phone, also known as a cellular phone, with more than 4.6 billion mobile cellular subscriptions worldwide as of the end of 2010. These wireless phones use radio waves from signal-transmission towers to enable their users to make phone calls from many locations worldwide. They can be used within range of the mobile telephone site used to house the equipment required to transmit and receive the radio signals from these instruments.

Wireless Data Communications

Wireless data communications are an essential component of mobile computing. The various available technologies differ in local availability, coverage range and performance, and in some circumstances, users must be able to employ multiple connection types and switch between them. To simplify the experience for the user, connection manager software can be used, or a mobile VPN deployed to handle the multiple connections as a secure, single virtual network. Supporting technologies include:

Wi-Fi is a wireless local area network that enables portable computing devices to connect easily to the Internet. Standardized as IEEE 802.11 a,b,g,n, Wi-Fi approaches speeds of some types of wired Ethernet. Wi-Fi has become the de facto standard for access in private homes, within offices, and at public hotspots. Some businesses charge customers a monthly fee for service, while others have begun offering it for free in an effort to increase the sales of their goods.

Cellular data service offers coverage within a range of 10-15 miles from the nearest cell site. Speeds have increased as technologies have evolved, from earlier technologies such as GSM, CDMA and GPRS, to 3G networks such as W-CDMA, EDGE or CDMA2000.

Mobile Satellite Communications may be used where other wireless connections are unavailable, such as in largely rural areas or remote locations. Satellite communications are especially important for transportation, aviation, maritime and military use.

Wireless Sensor Networks are responsible for sensing noise, interference, and activity in data collection networks. This allows us to detect relevant quantities, monitor and collect data, formulate clear user displays, and to perform decision-making functions.

Wireless Energy Transfer

Wireless energy transfer is a process whereby electrical energy is transmitted from a power source to an electrical load (Computer Load) that does not have a built-in power source, without the use of interconnecting wires. There are two different fundamental methods for wireless energy transfer. They can be transferred using either far-field methods that involve beaming power/lasers, radio or microwave transmissions or near-field using induction. Both methods utilize electromagnetism and magnetic fields.

Wireless Medical Technologies

New wireless technologies, such as mobile body area networks (MBAN), have the capability to monitor blood pressure, heart rate, oxygen level and body temperature. The MBAN works by sending low powered wireless signals to receivers that feed into nursing stations or monitoring sites. This technology helps with the intentional and unintentional risk of infection or disconnection that arise from wired connections.

Computer Interface Devices

Answering the call of customers frustrated with cord clutter, many manufacturers of computer peripherals turned to wireless technology to satisfy their consumer base. Originally these units used bulky, highly local transceivers to mediate between a computer and a keyboard and mouse; however, more recent generations have used small, high-quality devices, some even incorporating Bluetooth. These systems have become so ubiquitous that some users have begun complaining about a lack of wired peripherals. Wireless devices tend to have a slightly slower response time than their wired counterparts; however, the gap is decreasing.

A battery powers computer interface devices such as a keyboard or mouse and send signals to a receiver through a USB port by the way of a radio frequency (RF) receiver. The RF design makes it possible for signals to be transmitted wirelessly and expands the range of efficient use, usually up to 10 feet. Distance, physical obstacles, competing signals, and even human bodies can all degrade the signal quality.

Concerns about the security of wireless keyboards arose at the end of 2007, when it was revealed that Microsoft's implementation of encryption in some of its 27 MHz models was highly insecure.

Categories of Wireless Implementations, Devices and Standards

- Radio station in accordance with ITU RR.

- Radiocommunication service in accordance with ITU RR.

- Radio communication system.

- Land Mobile Radio or Professional Mobile Radio: TETRA, P25, OpenSky, EDACS, DMR, dPMR.

- Cordless telephony:DECT (Digital Enhanced Cordless Telecommunications).

- Cellular networks: 0G, 1G, 2G, 3G, Beyond 3G (4G), Future wireless.

- List of emerging technologies.

- Short-range point-to-point communication : Wireless microphones, Remote controls, IrDA, RFID (Radio Frequency Identification), TransferJet, Wireless USB, DSRC (Dedicated Short Range Communications), EnOcean, Near Field Communication.

- Wireless sensor networks: ZigBee, EnOcean; Personal area networks, Bluetooth, Transfer-Jet, Ultra-wideband (UWB from WiMedia Alliance).

- Wireless networks: Wireless LAN (WLAN), (IEEE 802.11 branded as Wi-Fi and Hiper-LAN), Wireless Metropolitan Area Networks (WMAN) and (LMDS, WiMAX, and Hiper-MAN).

Wireless Network

A wireless network is any type of computer network that uses wireless data connections for connecting network nodes.

Wireless networking is a method by which homes, telecommunications networks and enterprise (business) installations avoid the costly process of introducing cables into a building, or as a connection between various equipment locations. Wireless telecommunications networks are generally implemented and administered using radio communication. This implementation takes place at the physical level (layer) of the OSI model network structure.

Wireless icon.

Examples of wireless networks include cell phone networks, Wireless local networks, wireless sensor networks, satellite communication networks, and terrestrial microwave networks.

History

Wireless Links

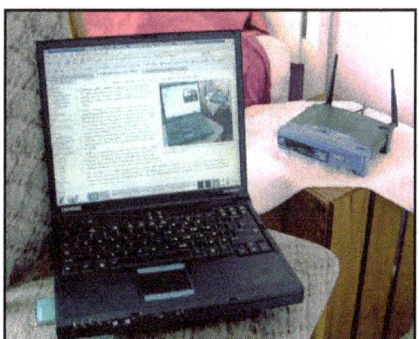
Computers are very often connected to networks using wireless links.

- *Terrestrial microwave* – Terrestrial microwave communication uses Earth-based transmitters and receivers resembling satellite dishes. Terrestrial microwaves are in the low gigahertz range, which limits all communications to line-of-sight. Relay stations are spaced approximately 48 km (30 mi) apart.

- *Communications satellites* – Satellites communicate via microwave radio waves, which are not deflected by the Earth's atmosphere. The satellites are stationed in space, typically in geosynchronous orbit 35,400 km (22,000 mi) above the equator. These Earth-orbiting systems are capable of receiving and relaying voice, data, and TV signals.

- *Cellular and PCS systems* use several radio communications technologies. The systems divide the region covered into multiple geographic areas. Each area has a low-power transmitter or radio relay antenna device to relay calls from one area to the next area.

- *Radio and spread spectrum technologies* – Wireless local area networks use a high-frequency radio technology similar to digital cellular and a low-frequency radio technology. Wireless LANs use spread spectrum technology to enable communication between multiple devices in a limited area. IEEE 802.11 defines a common flavor of open-standards wireless radio-wave technology known as Wifi.

- *Free-space optical communication* uses visible or invisible light for communications. In most cases, line-of-sight propagation is used, which limits the physical positioning of communicating devices.

Types of Wireless Networks

Wireless PAN

Wireless personal area networks (WPANs) interconnect devices within a relatively small area, that is generally within a person's reach. For example, both Bluetooth radio and invisible infrared light provides a WPAN for interconnecting a headset to a laptop. ZigBee also supports WPAN applications. Wi-Fi PANs are becoming commonplace (2010) as equipment designers start to integrate Wi-Fi into a variety of consumer electronic devices. Intel "My WiFi" and Windows 7 "virtual Wi-Fi" capabilities have made Wi-Fi PANs simpler and easier to set up and configure.

Wireless LAN

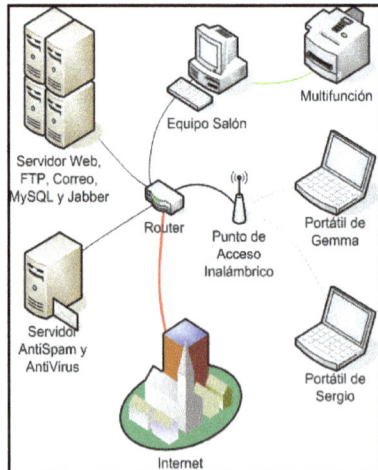

Wireless LANs are often used for connecting to local resources and to the Internet.

A wireless local area network (WLAN) links two or more devices over a short distance using a wireless distribution method, usually providing a connection through an access point for internet access. The use of spread-spectrum or OFDM technologies may allow users to move around within a local coverage area, and still remain connected to the network.

Products using the IEEE 802.11 WLAN standards are marketed under the Wi-Fi brand name. Fixed wireless technology implements point-to-point links between computers or networks at two distant locations, often using dedicated microwave or modulated laser light beams over line of sight paths. It is often used in cities to connect networks in two or more buildings without installing a wired link.

Wireless Mesh Network

A wireless mesh network is a wireless network made up of radio nodes organized in a mesh topology. Each node forwards messages on behalf of the other nodes. Mesh networks can "self-heal", automatically re-routing around a node that has lost power.

Wireless MAN

Wireless metropolitan area networks are a type of wireless network that connects several wireless LANs.

- WiMAX is a type of Wireless MAN and is described by the IEEE 802.16 standard.

Wireless WAN

Wireless wide area networks are wireless networks that typically cover large areas, such as between neighbouring towns and cities, or city and suburb. These networks can be used to connect branch offices of business or as a public Internet access system. The wireless connections between access points are usually point to point microwave links using parabolic dishes on the 2.4 GHz band, rather than omnidirectional antennas used with smaller networks. A typical system contains base station gateways, access points and wireless bridging relays. Other configurations are mesh systems where each access point acts as a relay also. When combined with renewable energy systems such as photovoltaic solar panels or wind systems they can be stand alone systems.

Global Area Network

A global area network (GAN) is a network used for supporting mobile across an arbitrary number of wireless LANs, satellite coverage areas, etc. The key challenge in mobile communications is handing off user communications from one local coverage area to the next. In IEEE Project 802, this involves a succession of terrestrial wireless LANs.

Space Network

Space networks are networks used for communication between spacecraft, usually in the vicinity of the Earth. The example of this is NASA's Space Network.

Different Uses

Some examples of usage include cellular phones which are part of everyday wireless networks, allowing easy personal communications. Another example, Intercontinental network systems, use radio satellites to communicate across the world. Emergency services such as the police utilize wireless networks to communicate effectively as well. Individuals and businesses use wireless networks to send and share data rapidly, whether it be in a small office building or across the world.

Properties

General

In a general sense, wireless networks offer a vast variety of uses by both business and home users.

"Now, the industry accepts a handful of different wireless technologies. Each wireless technology is defined by a standard that describes unique functions at both the Physical and the Data Link layers of the OSI model. These standards differ in their specified signaling methods, geographic ranges, and frequency usages, among other things. Such differences can make certain technologies better suited to home networks and others better suited to network larger organizations."

Performance

Each standard varies in geographical range, thus making one standard more ideal than the next depending on what it is one is trying to accomplish with a wireless network. The performance of wireless networks satisfies a variety of applications such as voice and video. The use of this technology also gives room for expansions, such as from 2G to 3G and, most recently, 4G technology, which stands for the fourth generation of cell phone mobile communications standards. As wireless networking has become commonplace, sophistication increases through configuration of network hardware and software, and greater capacity to send and receive larger amounts of data, faster, is achieved.

Space

Space is another characteristic of wireless networking. Wireless networks offer many advantages when it comes to difficult-to-wire areas trying to communicate such as across a street or river, a warehouse on the other side of the premises or buildings that are physically separated but operate as one. Wireless networks allow for users to designate a certain space which the network will be able to communicate with other devices through that network. Space is also created in homes as a result of eliminating clutters of wiring. This technology allows for an alternative to installing physical network mediums such as TPs, coaxes, or fiber-optics, which can also be expensive.

Home

For homeowners, wireless technology is an effective option compared to Ethernet for sharing printers, scanners, and high-speed Internet connections. WLANs help save the cost of installation of cable mediums, save time from physical installation, and also creates mobility for devices connected to the network. Wireless networks are simple and require as few as one single wireless access point connected directly to the Internet via a router.

Wireless Network Elements

The telecommunications network at the physical layer also consists of many interconnected wireline network elements (NEs). These NEs can be stand-alone systems or products that are either supplied by a single manufacturer or are assembled by the service provider (user) or system integrator with parts from several different manufacturers.

Wireless NEs are the products and devices used by a wireless carrier to provide support for the backhaul network as well as a mobile switching center (MSC).

Reliable wireless service depends on the network elements at the physical layer to be protected against all operational environments and applications.

What are especially important are the NEs that are located on the cell tower to the base station (BS) cabinet. The attachment hardware and the positioning of the antenna and associated closures and cables are required to have adequate strength, robustness, corrosion resistance, and resistance against wind, storms, icing, and other weather conditions. Requirements for individual components, such as hardware, cables, connectors, and closures, shall take into consideration the structure to which they are attached.

Difficulties

Interferences

Compared to wired systems, wireless networks are frequently subject to electromagnetic interference. This can be caused by other networks or other types of equipment that generate radio waves that are within, or close, to the radio bands used for communication. Interference can degrade the signal or cause the system to fail.

Absorption and Reflection

Some materials cause absorption of electromagnetic waves, preventing it from reaching the receiver, in other cases, particularly with metallic or conductive materials reflection occurs. This can cause dead zones where no reception is available. Aluminium foiled thermal isolation in modern homes can easily reduce indoor mobile signals by 10 dB frequently leading to complaints about the bad reception of long-distance rural cell signals.

Multipath Fading

In multipath fading two or more different routes taken by the signal, due to reflections, can cause the signal to cancel out at certain locations, and to be stronger in other places (upfade).

Hidden Node Problem

The hidden node problem occurs in some types of network when a node is visible from a wireless access point (AP), but not from other nodes communicating with that AP. This leads to difficulties in media access control.

Shared Resource Problem

The wireless spectrum is a limited resource and shared by all nodes in the range of its transmitters. Bandwidth allocation becomes complex with multiple participating users. Often users are not aware that advertised numbers (e.g., for IEEE 802.11 equipment or LTE networks) are not their capacity, but shared with all other users and thus the individual user rate is far lower. With increasing demand, the capacity crunch is more and more likely to happen. User-in-the-loop (UIL) may be an alternative solution to ever upgrading to newer technologies for over-provisioning.

Capacity

Channel

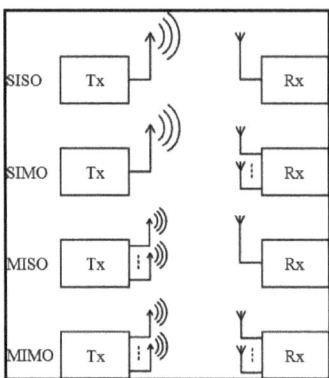

Understanding of SISO, SIMO, MISO and MIMO. Using multiple antennas and transmitting in different frequency channels can reduce fading, and can greatly increase the system capacity.

Shannon's theorem can describe the maximum data rate of any single wireless link, which relates to the bandwidth in hertz and to the noise on the channel.

One can greatly increase channel capacity by using MIMO techniques, where multiple aerials or multiple frequencies can exploit multiple paths to the receiver to achieve much higher throughput – by a factor of the product of the frequency and aerial diversity at each end.

Under Linux, the Central Regulatory Domain Agent (CRDA) controls the setting of channels.

Network

The total network bandwidth depends on how dispersive the medium is (more dispersive medium generally has better total bandwidth because it minimises interference), how many frequencies are available, how noisy those frequencies are, how many aerials are used and whether a directional antenna is in use, whether nodes employ power control and so on.

Cellular wireless networks generally have good capacity, due to their use of directional aerials, and their ability to reuse radio channels in non-adjacent cells. Additionally, cells can be made very small using low power transmitters this is used in cities to give network capacity that scales linearly with population density.

Safety

Wireless access points are also often close to humans, but the drop off in power over distance is fast, following the inverse-square law. The position of the United Kingdom's Health Protection Agency (HPA) is that "...radio frequency (RF) exposures from WiFi are likely to be lower than those from mobile phones." It also saw "...no reason why schools and others should not use WiFi equipment." In October 2007, the HPA launched a new "systematic" study into the effects of WiFi networks on behalf of the UK government, in order to calm fears that had appeared in the media in a recent period up to that time". Dr Michael Clark, of the HPA, says published research on mobile phones and masts does not add up to an indictment of WiFi.

Wireless LAN

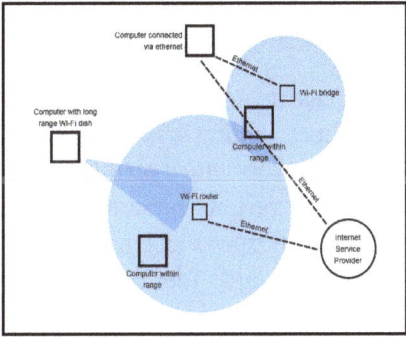

An example of a Wi-Fi network.

A wireless local area network (WLAN) is a wireless computer network that links two or more devices using a wireless distribution method (often spread-spectrum or OFDM radio) within a limited area such as a home, school, computer laboratory, or office building. This gives users the ability to move around within a local coverage area and still be connected to the network, and can provide a connection to the wider Internet. Most modern WLANs are based on IEEE 802.11 standards, marketed under the Wi-Fi brand name.

Wireless LANs have become popular in the home due to ease of installation and use, and in commercial complexes offering wireless access to their customers; often for free. New York City, for instance, has begun a pilot program to provide city workers in all five boroughs of the city with wireless Internet access.

An embedded RouterBoard 112 with U.FL-RSMA pigtail and R52 mini PCI Wi-Fi card widely used by wireless Internet service providers (WISPs).

History

Norman Abramson, a professor at the University of Hawaii, developed the world's first wireless computer communication network, ALOHAnet (operational in 1971), using low-cost ham-like radios. The system included seven computers deployed over four islands to communicate with the central computer on the Oahu Island without using phone lines.

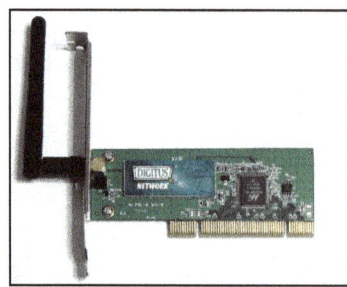

54 Mbit/s WLAN PCI Card (802.11g).

WLAN (Wireless Local Area Network) hardware initially cost so much that it was only used as an alternative to cabled LAN in places where cabling was difficult or impossible. Early development included industry-specific solutions and proprietary protocols, but at the end of the 1990s these were replaced by standards, primarily the various versions of IEEE 802.11 (in products using the Wi-Fi brand name). Beginning in 1991, a European alternative known as HiperLAN/1 was pursued by the European Telecommunications Standards Institute (ETSI) with a first version approved in 1996. This was followed by a HiperLAN/2 functional specification with ATM influences accomplished February 2000. Neither European standard achieved the commercial success of 802.11, although much of the work on HiperLAN/2 has survived in the PHY specification for IEEE 802.11a, which is nearly identical to the PHY of HiperLAN/2.

In 2009 802.11n was added to 802.11. It operates in both the 2.4 GHz and 5 GHz bands at a maximum data transfer rate of 600 Mbit/s. Most newer routers are able to utilise both wireless bands, known as dualband. This allows data communications to avoid the crowded 2.4 GHz band, which is also shared with Bluetooth devices and microwave ovens. The 5 GHz band is also wider than the 2.4 GHz band, with more channels, which permits a greater number of devices to share the space. Not all channels are available in all regions.

A HomeRF group formed in 1997 to promote a technology aimed for residential use, but it disbanded at the end of 2002.

Architecture

Stations

All components that can connect into a wireless medium in a network are referred to as stations (STA). All stations are equipped with wireless network interface controllers (WNICs). Wireless stations fall into one of two categories: wireless access points, and clients. Access points (APs), normally wireless routers, are base stations for the wireless network. They transmit and receive radio frequencies for wireless enabled devices to communicate with. Wireless clients can be mobile devices such as laptops, personal digital assistants, IP phones and other smartphones, or fixed devices such as desktops and workstations that are equipped with a wireless network interface.

Basic Service Set

The basic service set (BSS) is a set of all stations that can communicate with each other at PHY layer. Every BSS has an identification (ID) called the BSSID, which is the MAC address of the access point servicing the BSS.

There are two types of BSS: Independent BSS (also referred to as IBSS), and infrastructure BSS. An independent BSS (IBSS) is an ad hoc network that contains no access points, which means they cannot connect to any other basic service set.

Extended Service Set

An extended service set (ESS) is a set of connected BSSs. Access points in an ESS are connected by a distribution system. Each ESS has an ID called the SSID which is a 32-byte (maximum) character string.

Distribution System

A distribution system (DS) connects access points in an extended service set. The concept of a DS can be used to increase network coverage through roaming between cells.

DS can be wired or wireless. Current wireless distribution systems are mostly based on WDS or MESH protocols, though other systems are in use.

Types of Wireless Lans

The IEEE 802.11 has two basic modes of operation: infrastructure and **ad hoc** mode. In *ad hoc* mode, mobile units transmit directly peer-to-peer. In infrastructure mode, mobile units communicate through an access point that serves as a bridge to other networks (such as Internet or LAN).

Since wireless communication uses a more open medium for communication in comparison to wired LANs, the 802.11 designers also included encryption mechanisms: Wired Equivalent Privacy (WEP, now insecure), Wi-Fi Protected Access (WPA, WPA2), to secure wireless computer networks. Many access points will also offer Wi-Fi Protected Setup, a quick (but now insecure) method of joining a new device to an encrypted network.

Infrastructure

Most Wi-Fi networks are deployed in infrastructure mode.

In infrastructure mode, a base station acts as a wireless access point hub, and nodes communicate through the hub. The hub usually, but not always, has a wired or fiber network connection, and may have permanent wireless connections to other nodes.

Wireless access points are usually fixed, and provide service to their client nodes within range.

Wireless clients, such as laptops, smartphones etc. connect to the access point to join the network.

Sometimes a network will have a multiple access points, with the same 'SSID' and security arrangement. In that case connecting to any access point on that network joins the client to the network. In that case, the client software will try to choose the access point to try to give the best service, such as the access point with the strongest signal.

Peer-to-peer

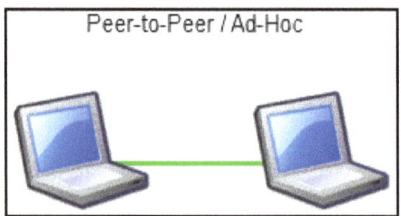

Peer-to-Peer or ad hoc wireless LAN.

An ad hoc network (not the same as a WiFi Direct network) is a network where stations communicate only peer to peer (P2P). There is no base and no one gives permission to talk. This is accomplished using the Independent Basic Service Set (IBSS).

A WiFi Direct network is another type of network where stations communicate peer to peer.

In a Wi-Fi P2P group, the group owner operates as an access point and all other devices are clients. There are two main methods to establish a group owner in the Wi-Fi Direct group. In one approach, the user sets up a P2P group owner manually. This method is also known as Autonomous Group Owner (autonomous GO). In the second method, also called negotiation-based group creation, two devices compete based on the group owner intent value. The device with higher intent value becomes a group owner and the second device becomes a client. Group owner intent value can depend on whether the wireless device performs a cross-connection between an infrastructure WLAN service and a P2P group, remaining power in the wireless device, whether the wireless device is already a group owner in another group and/or a received signal strength of the first wireless device.

A peer-to-peer network allows wireless devices to directly communicate with each other. Wireless devices within range of each other can discover and communicate directly without involving central access points. This method is typically used by two computers so that they can connect to each other to form a network. This can basically occur in devices within a closed range.

If a signal strength meter is used in this situation, it may not read the strength accurately and can be misleading, because it registers the strength of the strongest signal, which may be the closest computer.

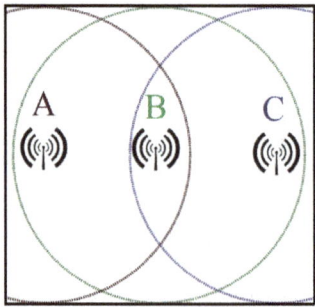

Hidden node problem: Devices A and C are both communicating with B, but are unaware of each other.

IEEE 802.11 defines the physical layer (PHY) and MAC (Media Access Control) layers based on CSMA/CA (Carrier Sense Multiple Access with Collision Avoidance). The 802.11 specification includes provisions designed to minimize collisions, because two mobile units may both be in range of a common access point, but out of range of each other.

Bridge

''*A bridge can be used to connect networks, typically of different types. A wireless Ethernet bridge allows the connection of devices on a wired Ethernet network to a wireless network. The bridge acts as the connection point to the Wireless LAN.*

Wireless Distribution System

A Wireless Distribution System enables the wireless interconnection of access points in an IEEE 802.11 network. It allows a wireless network to be expanded using multiple access points without the need for a wired backbone to link them, as is traditionally required. The notable advantage of WDS over other solutions is that it preserves the MAC addresses of client packets across links between access points.

An access point can be either a main, relay or remote base station. A main base station is typically connected to the wired Ethernet. A relay base station relays data between remote base stations, wireless clients or other relay stations to either a main or another relay base station. A remote base station accepts connections from wireless clients and passes them to relay or main stations. Connections between "clients" are made using MAC addresses rather than by specifying IP assignments.

All base stations in a Wireless Distribution System must be configured to use the same radio channel, and share WEP keys or WPA keys if they are used. They can be configured to different service set identifiers. WDS also requires that every base station be configured to forward to others in the system as mentioned above.

WDS may also be referred to as repeater mode because it appears to bridge and accept wireless clients at the same time (unlike traditional bridging). It should be noted, however, that throughput in this method is halved for all clients connected wirelessly.

When it is difficult to connect all of the access points in a network by wires, it is also possible to put up access points as repeaters.

Roaming

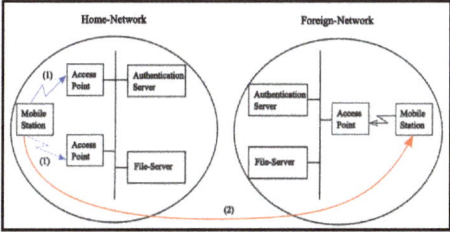

Roaming among Wireless Local Area Networks.

There are two definitions for wireless LAN roaming:

- Internal Roaming: The Mobile Station (MS) moves from one access point (AP) to another AP within a home network because the signal strength is too weak. An authentication server (RADIUS) performs the re-authentication of MS via 802.1x (e.g. with PEAP). The billing of QoS is in the home network. A Mobile Station roaming from one access point to another often interrupts the flow of data among the Mobile Station and an application connected to the network. The Mobile Station, for instance, periodically monitors the presence of alternative access points (ones that will provide a better connection). At some point, based on proprietary mechanisms, the Mobile Station decides to re-associate with an access point having a stronger wireless signal. The Mobile Station, however, may lose a connection with an access point before associating with another access point. In order to provide reliable connections with applications, the Mobile Station must generally include software that provides session persistence.

- External Roaming: The MS (client) moves into a WLAN of another Wireless Internet Service Provider (WISP) and takes their services (Hotspot). The user can independently of his home network use another foreign network, if this is open for visitors. There must be special authentication and billing systems for mobile services in a foreign network.

Applications

Wireless LANs have a great deal of applications. Modern implementations of WLANs range from small in-home networks to large, campus-sized ones to completely mobile networks on airplanes and trains.

Users can access the Internet from WLAN hotspots in restaurants, hotels, and now with portable devices that connect to 3G or 4G networks. Oftentimes these types of public access points require no registration or password to join the network. Others can be accessed once registration has occurred and/or a fee is paid.

Existing Wireless LAN infrastructures can also be used to work as indoor positioning systems with no modification to the existing hardware.

Performance and Throughput

WLAN, organised in various layer 2 variants (IEEE 802.11), has different characteristics. Across all flavours of 802.11, maximum achievable throughputs are either given based on measurements under ideal conditions or in the layer 2 data rates. This, however, does not apply to typical deployments in which data are being transferred between two endpoints of which at least one is typically connected to a wired infrastructure and the other endpoint is connected to an infrastructure via a wireless link.

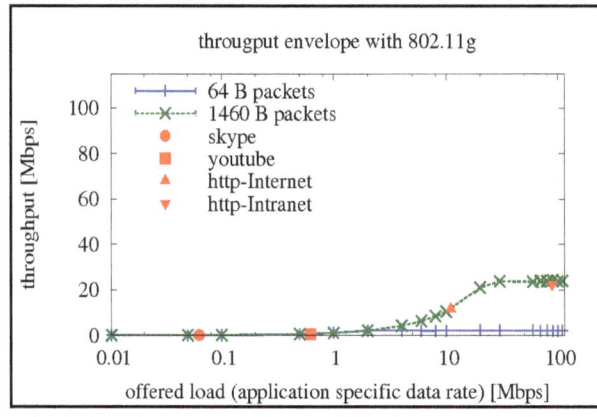

Graphical representation of Wi-Fi application specific (UDP) performance envelope 2.4 GHz band, with 802.11g.

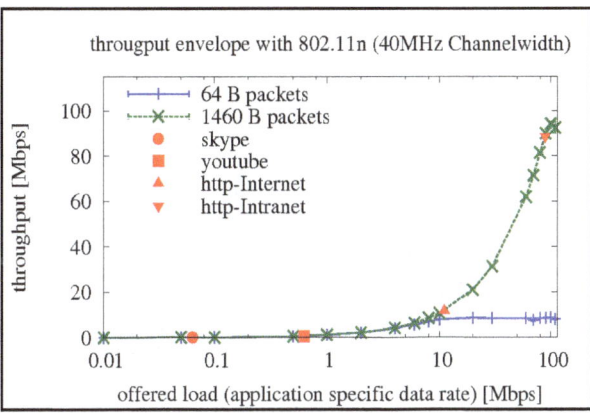

Graphical representation of Wi-Fi application specific (UDP) performance envelope 2.4 GHz band, with 802.11n with 40 MHz.

This means that typically data frames pass an 802.11 (WLAN) medium and are being converted to 802.3 (Ethernet) or vice versa.

Due to the difference in the frame (header) lengths of these two media, the packet size of an application determines the speed of the data transfer. This means that an application which uses small packets (e.g. VoIP) creates a data flow with a high overhead traffic (e.g. a low goodput).

Other factors which contribute to the overall application data rate are the speed with which the application transmits the packets (i.e. the data rate) and the energy with which the wireless signal is received.

The latter is determined by distance and by the configured output power of the communicating devices.

Same references apply to the attached throughput graphs which show measurements of UDP throughput measurements. Each represents an average (UDP) throughput (the error bars are there, but barely visible due to the small variation) of 25 measurements.

Each is with a specific packet size (small or large) and with a specific data rate (10 kbit/s – 100 Mbit/s). Markers for traffic profiles of common applications are included as well. This text and measurements do not cover packet errors but information about this can be found at above references. The table below shows the maximum achievable (application specific) UDP throughput in the same scenarios (same references again) with various difference WLAN (802.11) flavours. The measurement hosts have been 25 meters apart from each other; loss is again ignored.

Wireless Access Point

In computer networking, a wireless access point (WAP) is a networking hardware device that allows a Wi-Fi compliant device to connect to a wired network. The WAP usually connects to a router (via a wired network) as a standalone device, but it can also be an integral component of the router itself. A WAP is differentiated from a hotspot, which is the physical location where Wi-Fi access to a WLAN is available.

Introduction

Linksys "WAP54G" 802.11g wireless access point.

Prior to wireless networks, setting up a computer network in a business, home or school often required running many cables through walls and ceilings in order to deliver network access to all of the network-enabled devices in the building. With the creation of the wireless access point, network users are now able to add devices that access the network with few or no cables. A WAP normally connects directly to a wired Ethernet connection and the WAP then provides wireless connections using radio frequency links for other devices to utilize that wired connection. Most WAPs support the connection of multiple wireless devices to one wired connection. Modern WAPs are built to support a standard for sending and receiving data using these radio frequencies. Those standards, and the frequencies they use are defined by the IEEE. Most APs use IEEE 802.11 standards.

Common AP Applications

Typical corporate use involves attaching several WAPs to a wired network and then providing wireless access to the office LAN. The wireless access points are managed by a WLAN Controller which handles automatic adjustments to RF power, channels, authentication, and security. Furthermore, controllers can be combined to form a wireless mobility group to allow inter-controller roaming. The controllers can be part of a mobility domain to allow clients access throughout large or regional office locations. This saves the clients time and administrators overhead because it can automatically re-associate or re-authenticate.

A hotspot is a common public application of WAPs, where wireless clients can connect to the Internet without regard for the particular networks to which they have attached for the moment. The concept has become common in large cities, where a combination of coffeehouses, libraries, as well as privately owned open access points, allow clients to stay more or less continuously connected to the Internet, while moving around. A collection of connected hotspots can be referred to as a lily pad network.

WAPs are commonly used in home wireless networks. Home networks generally have only one AP to connect all the computers in a home. Most are wireless routers, meaning converged devices that include the WAP, a router, and, often, an Ethernet switch. Many also include a broadband modem. In places where most homes have their own WAP within range of the neighbours' AP, it's possible for technically savvy people to turn off their encryption and set up a wireless community network, creating an intra-city communication network although this does not negate the requirement for a wired network.

A WAP may also act as the network's arbitrator, negotiating when each nearby client device can transmit. However, the vast majority of currently installed IEEE 802.11 networks do not implement this, using a distributed pseudo-random algorithm called CSMA/CA instead.

Wireless Access Point vs. Ad Hoc Network

Some people confuse wireless access points with wireless ad hoc networks. An ad hoc network uses a connection between two or more devices without using a wireless access point: the devices communicate directly when in range. An ad hoc network is used in situations such as a quick data exchange or a multiplayer LAN game because setup is easy and does not require an access point. Due to its peer-to-peer layout, ad hoc connections are similar to Bluetooth ones.

But ad hoc connections are generally not recommended for a permanent installation. The reason is that Internet access via ad hoc networks, using features like Windows' Internet Connection Sharing, may work well with a small number of devices that are close to each other, but ad hoc networks don't scale well. Internet traffic will converge to the nodes with direct internet connection, potentially congesting these nodes. For internet-enabled nodes, access points have a clear advantage, with the possibility of having multiple access points connected by a wired LAN.

Limitations

One IEEE 802.11 AP can typically communicate with 30 client systems located within a radius of 103 metres. However, the actual range of communication can vary significantly, depending on such variables as indoor or outdoor placement, height above ground, nearby obstructions, other electronic devices that might actively interfere with the signal by broadcasting on the same frequency, type of antenna, the current weather, operating radio frequency, and the power output of devices. Network designers can extend the range of APs through the use of repeaters and reflectors, which can bounce or amplify radio signals that ordinarily would go un-received. In experimental conditions, wireless networking has operated over distances of several hundred kilometers.

Most jurisdictions have only a limited number of frequencies legally available for use by wireless networks. Usually, adjacent WAPs will use different frequencies (Channels) to communicate with their clients in order to avoid interference between the two nearby systems. Wireless devices can "listen" for data traffic on other frequencies, and can rapidly switch from one frequency to another to achieve better reception. However, the limited number of frequencies becomes problematic in crowded downtown areas with tall buildings using multiple WAPs. In such an environment, signal overlap becomes an issue causing interference, which results in signal droppage and data errors.

Wireless networking lags wired networking in terms of increasing bandwidth and throughput. While (as of 2013) high-density 256-QAM (TurboQAM) modulation, 3-antenna wireless devices for the consumer market can reach sustained real-world speeds of some 240 Mbit/s at 13 m behind two standing walls (NLOS) depending on their nature or 360 Mbit/s at 10 m line of sight or 380 Mbit/s at 2 m line of sight (IEEE 802.11ac) or 20 to 25 Mbit/s at 2 m line of sight (IEEE 802.11g), wired hardware of similar cost reaches somewhat less than 1000 Mbit/s up to specified distance of 100 m with twisted-pair cabling (Cat5, Cat5e, Cat6, or Cat7) (Gigabit Ethernet). One impediment to increasing the speed of wireless communications comes from Wi-Fi's use of a shared communications medium: Thus, two stations in infrastructure mode that are communicating with each other even over the same AP must have each and every frame transmitted twice: from the sender to the AP, then from the AP to the receiver. This approximately halves the effective bandwidth, so an AP is only able to use somewhat less than half the actual over-the-air rate for data throughput. Thus a typical 54 Mbit/s wireless connection actually carries TCP/IP data at 20 to 25 Mbit/s. Users of legacy wired networks expect faster speeds, and people using wireless connections keenly want to the wireless networks catch up.

By 2012, 802.11n based access points and client devices have already taken a fair share of the marketplace and with the finalization of the 802.11n standard in 2009 inherent problems integrating products from different vendors are less prevalent.

Security

Wireless access has special security considerations. Many wired networks base the security on physical access control, trusting all the users on the local network, but if wireless access points are connected to the network, anybody within range of the AP (which typically extends farther than the intended area) can attach to the network.

The most common solution is wireless traffic encryption. Modern access points come with built-in encryption. The first generation encryption scheme 'WEP' proved easy to crack; the second and third generation schemes, WPA and WPA2, are considered secure if a strong enough password or passphrase is used.

Some APs support hotspot style authentication using RADIUS and other authentication servers.

Opinions about wireless network security vary widely. Bruce Schneier asserted the net benefits of open Wi-Fi without passwords outweigh the risks, a position supported in 2014 by Peter Eckersley of the Electronic Frontier Foundation.

The opposite position was taken by Nick Mediati in an article for *PC World*, in which he takes the position that every wireless access point should be locked down with a password.

Wireless WAN

A wireless wide area network (WWAN), is a form of wireless network. The larger size of a wide area network compared to a local area network requires differences in technology. Wireless networks of all sizes deliver data in the form of telephone calls, web pages, and streaming video.

A WWAN often differs from wireless local area network (WLAN) by using mobile telecommunication cellular network technologies such as LTE, WiMAX (often called a wireless metropolitan area network or WMAN), UMTS, CDMA2000, GSM, cellular digital packet data (CDPD) and Mobitex to transfer data. It can also use Local Multipoint Distribution Service (LMDS) or Wi-Fi to provide Internet access. These technologies are offered regionally, nationwide, or even globally and are provided by a wireless service provider. WWAN connectivity allows a user with a laptop and a WWAN card to surf the web, check email, or connect to a virtual private network (VPN) from anywhere within the regional boundaries of cellular service. Various computers can have integrated WWAN capabilities.

A WWAN may also be a closed network that covers a large geographic area. For example, a mesh network or MANET with nodes on building, tower, trucks, and planes could also considered a WWAN.

Since radio communications systems do not provide a physically secure connection path, WWANs typically incorporate encryption and authentication methods to make them more secure. Unfortunately some of the early GSM encryption techniques were flawed, and security experts have issued warnings that cellular communication, including WWAN, is no longer secure. UMTS (3G) encryption was developed later and has yet to be broken.

Wireless Mesh Network

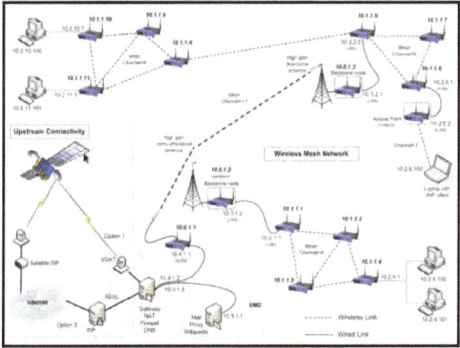

Diagram showing a possible configuration for a wireless mesh network, connected upstream via a VSAT link (click to enlarge).

A wireless mesh network (WMN) is a communications network made up of radio nodes organized in a mesh topology. It is also a form of wireless ad hoc network. Wireless mesh networks often consist of mesh clients, mesh routers and gateways. The mesh clients are often laptops, cell phones and other wireless devices while the mesh routers forward traffic to and from the gateways which may, but need not, be connected to the Internet. The coverage area of the radio nodes working as a single network is sometimes called a mesh cloud. Access to this mesh cloud is dependent on the radio nodes working in harmony with each other to create a radio network. A mesh network is reliable and offers redundancy. When one node can no longer operate, the rest of the nodes can still communicate with each other, directly or through one or more intermediate nodes. Wireless mesh networks can self form and self heal. Wireless mesh networks can be implemented with various wireless technologies including 802.11, 802.15, 802.16, cellular technologies and need not be restricted to any one technology or protocol.

History

Architecture

Wireless mesh architecture is a first step towards providing cost effective and dynamic high-bandwidth networks over a specific coverage area. Wireless mesh infrastructure is, in effect, a network of routers minus the cabling between nodes. It's built of peer radio devices that don't have to be cabled to a wired port like traditional WLAN access points (AP) do. Mesh infrastructure carries data over large distances by splitting the distance into a series of short hops. Intermediate nodes not only boost the signal, but cooperatively pass data from point A to point B by making forwarding decisions based on their knowledge of the network, i.e. perform routing. Such an architecture may, with careful design, provide high bandwidth, spectral efficiency, and economic advantage over the coverage area.

Wireless mesh networks have a relatively stable topology except for the occasional failure of nodes or addition of new nodes. The path of traffic, being aggregated from a large number of end users, changes infrequently. Practically all the traffic in an infrastructure mesh network is either forwarded to or from a gateway, while in ad hoc networks or client mesh networks the traffic flows between arbitrary pairs of nodes.

Management

This type of infrastructure can be decentralized (with no central server) or centrally managed (with a central server). Both are relatively inexpensive, and can be very reliable and resilient, as each node needs only transmit as far as the next node. Nodes act as routers to transmit data from nearby nodes to peers that are too far away to reach in a single hop, resulting in a network that can span larger distances. The topology of a mesh network is also reliable, as each node is connected to several other nodes. If one node drops out of the network, due to hardware failure or any other reason, its neighbors can quickly find another route using a routing protocol.

Applications

Mesh networks may involve either fixed or mobile devices. The solutions are as diverse as communication needs, for example in difficult environments such as emergency situations, tunnels, oil rigs, battlefield surveillance, high-speed mobile-video applications on board public transport or real-time racing-car telemetry. An important possible application for wireless mesh networks is VoIP. By using a Quality of Service scheme, the wireless mesh may support local telephone calls to be routed through the mesh.

Some current applications:

- U.S. military forces are now using wireless mesh networking to connect their computers, mainly ruggedized laptops, in field operations.

- Electric meters now being deployed on residences transfer their readings from one to another and eventually to the central office for billing without the need for human meter readers or the need to connect the meters with cables.

- The laptops in the One Laptop per Child program use wireless mesh networking to enable students to exchange files and get on the Internet even though they lack wired or cell phone or other physical connections in their area.

- The 66-satellite Iridium constellation operates as a mesh network, with wireless links between adjacent satellites. Calls between two satellite phones are routed through the mesh, from one satellite to another across the constellation, without having to go through an earth station. This makes for a smaller travel distance for the signal, reducing latency, and also allows for the constellation to operate with far fewer earth stations than would be required for 66 traditional communications satellites.

Operation

The principle is similar to the way packets travel around the wired Internet – data will hop from one device to another until it eventually reaches its destination. Dynamic routing algorithms implemented in each device allow this to happen. To implement such dynamic routing protocols, each device needs to communicate routing information to other devices in the network. Each device then determines what to do with the data it receives – either pass it on to the next device or keep it, depending on the protocol. The routing algorithm used should attempt to always ensure that the data takes the most appropriate (fastest) route to its destination.

Multi-radio Mesh

Multi-radio mesh refers to a unique pair of dedicated radios on each end of the link. This means there is a unique frequency used for each wireless hop and thus a dedicated CSMA collision domain. This is a true mesh link where you can achieve maximum performance without bandwidth degradation in the mesh and without adding latency. Thus voice and video applications work just as they would on a wired Ethernet network. In true 802.11 networks, there is no concept of a mesh. There are only APs and Stations. A multi-radio wireless mesh node will dedicate one of the radios to act as a station, and connect to a neighbor node AP radio.

Research Topics

One of the more often cited papers on Wireless Mesh Networks identified the following areas as open research problems in 2005:

- New modulation scheme:
    - In order to achieve higher transmission rate, new wideband transmission schemes other than OFDM and UWB are needed.
- Advanced antenna processing:
    - Advanced antenna processing including directional, smart and multiple antenna technologies is further investigated, since their complexity and cost are still too high for wide commercialization.
- Flexible spectrum management:
    - Tremendous efforts on research of frequency-agile techniques are being performed for increased efficiency.
- Cross-layer optimization:
    - Cross-layer research is a popular current research topic where information is shared between different communications layers in order to increase the knowledge and

current state of the network. This could enable new and more efficient protocols to be developed. A joint protocol which combines various design problems like routing, scheduling, channel assignment etc. can achieve higher performance since it is proven that these problems are strongly co-related. It is important to note that careless cross-layer design could lead to code which is difficult to maintain and extend.

- Software-defined wireless networking:

  ◦ Centralized, distributed, or hybrid? - In a new SDN architecture for WDNs is explored that eliminates the need for multi-hop flooding of route information and therefore enables WDNs to easily expand. The key idea is to split network control and data forwarding by using two separate frequency bands. The forwarding nodes and the SDN controller exchange link-state information and other network control signaling in one of the bands, while actual data forwarding takes place in the other band.

Protocols

Routing Protocols

There are more than 70 competing schemes for routing packets across mesh networks. Some of these include:

- AODV (Ad hoc On-Demand Distance Vector).
- B.A.T.M.A.N. (Better Approach To Mobile Adhoc Networking).
- Babel (protocol) (a distance-vector routing protocol for IPv6 and IPv4 with fast convergence properties).
- DNVR (Dynamic NIx-Vector Routing).
- DSDV (Destination-Sequenced Distance-Vector Routing).
- DSR (Dynamic Source Routing).
- HSLS (Hazy-Sighted Link State).
- HWMP (Hybrid Wireless Mesh Protocol).
- IWMP (Infrastructure Wireless Mesh Protocol) for Infrastructure Mesh Networks by GRECO UFPB-Brazil.
- Wireless mesh networks routing protocol (MRP) by Jangeun Jun and Mihail L. Sichitiu.
- OLSR (Optimized Link State Routing protocol).
- OORP (OrderOne Routing Protocol) (OrderOne Networks Routing Protocol).
- OSPF (Open Shortest Path First Routing).
- Routing Protocol for Low-Power and Lossy Networks (IETF ROLL RPL protocol, RFC 6550).

- PWRP (Predictive Wireless Routing Protocol).

- TORA (Temporally-Ordered Routing Algorithm).

- ZRP (Zone Routing Protocol).

The IEEE is developing a set of standards under the title 802.11s to define an architecture and protocol for ESS Mesh Networking.

A less thorough list can be found at Ad hoc routing protocol list.

Autoconfiguration Protocols

Standard autoconfiguration protocols, such as DHCP or IPv6 stateless autoconfiguration may be used over mesh networks.

Mesh network specific autoconfiguration protocols include:

- Ad Hoc Configuration Protocol (AHCP).

- Proactive Autoconfiguration (Proactive Autoconfiguration Protocol).

- Dynamic WMN Configuration Protocol (DWCP).

Communities and Providers

- CUWiN,

- Freifunk (DE) / FunkFeuer (AT) / OpenWireless (CH),

- Firetide,

- Guifi.net,

- Netsukuku,

- Ninux (IT),

- Senceive.

Hotspot

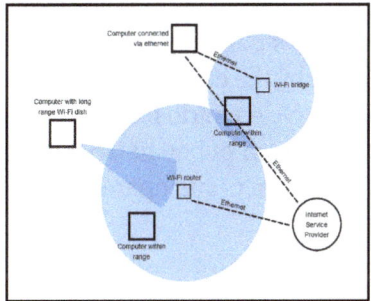

A diagram showing a Wi-Fi network.

A hotspot is a physical location where people may obtain Internet access, typically using Wi-Fi technology, via a wireless local area network (WLAN) using a router connected to an internet service provider.

Public hotspots may be found in an increasing number of businesses for use of customers in many developed urban areas throughout the world, such as coffee shops. Many hotels offer wifi access to guests, either in guest rooms or in the lobby. Hotspots differ from wireless access points, which are the hardware devices used to provide a wireless network service. Private hotspots allow Internet access to a device (such as a tablet) via another device which may have data access via say a mobile device.

History

Public park in Brooklyn, New York, has free Wi-Fi from a local corporation.

Public access wireless local area networks (LANs) were first proposed by Henrik Sjödin at the NetWorld+Interop conference in The Moscone Center in San Francisco in August 1993. Sjödin did not use the term hotspot but referred to publicly accessible wireless LANs.

The first commercial venture to attempt to create a public local area access network was a firm founded in Richardson, Texas known as PLANCOM (Public Local Area Network Communications). The founders of the venture, Mark Goode, Greg Jackson, and Brett Stewart dissolved the firm in 1998, while Goode and Jackson created MobileStar Networks. The firm was one of the first to sign such public access locations as Starbucks, American Airlines, and Hilton Hotels. The company was sold to Deutsche Telecom in 2001, who then converted the name of the firm into "T-Mobile Hotspot." It was then that the term "hotspot" entered the popular vernacular as a reference to a location where a publicly accessible wireless LAN is available.

ABI Research reported there was a total of 4.9 million global Wi-Fi hotspots in 2012 and projected that number would surpass 6.3 million by the end of 2013. The latest Wireless Broadband Alliance (WBA) Industry Report outlines a positive scenario for the Wi-Fi market: a steady annual increase from 5.2m public hotspots in 2012 to 10.5m public hotspots in 2018. Collectively, WBA operator members serve more than 1 billion subscribers and operate more than 15 million hotspots globally.

Uses

The public can use a laptop or other suitable portable device to access the wireless connection (usually Wi-Fi) provided. Of the estimated 150 million laptops, 14 million PDAs, and other emerging Wi-Fi devices sold per year for the last few years, most include the Wi-Fi feature.

For venues that have broadband Internet access, offering wireless access is as simple as configuring one access point (AP), in conjunction with a router and connecting the AP to the Internet connection. A single wireless router combining these functions may suffice.

The iPass 2014 interactive map, that shows data provided by the analysts Maravedis Rethink, shows that in December 2014 there are 46,000,000 hotspots worldwide and more than 22,000,000 roamable hotspots. More than 10,900 hotspots are on trains, planes and airports (Wi-Fi in motion) and more than 8,500,000 are "branded" hotspots (retail, cafés, hotels). The region with the largest number of public hotspots is Europe, followed by North America and Asia.

Security

Security is a serious concern in connection with Hotspots. There are three possible attack vectors. First, there is the wireless connection between the client and the access point. This needs to be encrypted, so that the connection cannot be eavesdropped or attacked by a man-in-the-middle-attack. Second, there is the Hotspot itself. The WLAN encryption ends at the interface, then travels its network stack unencrypted and then travels over the wired connection up to the BRAS of the ISP. Third, there is the connection from the Access Point to the BRAS of the ISP.

The safest method when accessing the Internet over a Hotspot, with unknown security measures, is end-to-end encryption. Examples of strong end-to-end encryption are HTTPS and SSH.

Locations

Hotspots are often found at airports, bookstores, coffee shops, department stores, fuel stations, hotels, hospitals, libraries, public pay phones, restaurants, RV parks and campgrounds, supermarkets, train stations, and other public places. Additionally, many schools and universities have wireless networks in their campuses.

Types

Free hotspots operate in two ways:

- Using an open public network is the easiest way to create a free hotspot. All that is needed is a Wi-Fi router. Similarly, when users of private wireless routers turn off their authentication requirements, opening their connection, intentionally or not, they permit piggybacking (sharing) by anyone in range.

- Closed public networks use a HotSpot Management System to control access to hotspots. This software runs on the router itself or an external computer allowing operators to authorize only specific users to access the Internet. Providers of such hotspots often associate the free access with a menu, membership, or purchase limit. Operators may also limit each user's available bandwidth (upload and download speed) to ensure that everyone gets a good quality service. Often this is done through service-level agreements.

Commercial Hotspots

A commercial hotspot may feature:

- A captive portal / login screen / splash page that users are redirected to for authentication and/or payment. The captive portal / splash page sometimes includes the social login buttons.

- A payment option using a credit card, iPass, PayPal, or another payment service (voucher-based Wi-Fi).

- A walled garden feature that allows free access to certain sites.

- Service-oriented provisioning to allow for improved revenue.

- Data analytics and data capture tools, to analyze and export data from Wi-Fi clients.

Many services provide payment services to hotspot providers, for a monthly fee or commission from the end-user income. For example, Amazingports can be used to set up hotspots that intend to offer both fee-based and free internet access, and ZoneCD is a Linux distribution that provides payment services for hotspot providers who wish to deploy their own service.

Major airports and business hotels are more likely to charge for service, though most hotels provide free service to guests; and increasingly, small airports and airline lounges offer free service.. Retail shops, public venues and offices usually provide a free Wi-Fi SSID for their guests and visitors.

Roaming services are expanding among major hotspot service providers. With roaming service the users of a commercial provider can have access to other providers' hotspots, either free of charge or for extra fees, which users will usually be charged on an access-per-minute basis.

Software Hotspots

Many Wi-Fi adapters built into or easily added to consumer computers and mobile devices include the functionality to operate as private or mobile hotspots, sometimes referred to as "mi-fi". The use of a private hotspot to enable other personal devices to access the WAN (usually but not always the Internet) is a form of bridging, and known as tethering. Manufacturers and firmware creators can enable this functionality in Wi-Fi devices on many Wi-Fi devices, depending upon the capabilities of the hardware, and most modern consumer operating systems, including Android, Apple OS X 10.6 and later, Windows mobile, and Linux include features to support this. Additionally wireless chipset manufacturers such as Atheros, Broadcom, Intel and others, may add the capability for certain Wi-Fi NICs, usually used in a client role, to also be used for hotspot purposes. However, some service providers, such as AT&T, Sprint, and T-Mobile charge users for this service or prohibit and disconnect user connections if tethering is detected.

Third-party software vendors offer applications to allow users to operate their own hotspot, whether to access the Internet when on the go, share an existing connection, or extend the range of another hotspot. Third party implementations of software hotspots include:

- AmazingPorts Hotspot software.

- Antamedia HotSpot software.
- Connectify Hotspot.
- Jaze Hotspot Gateway by Jaze Networks.
- Hot Spot Network Manager (HSNM).
- Virtual Router.
- Tanaza.
- Start Hotspot software.

Hotspot 2.0

Hotspot 2.0, also known as HS2 and Wi-Fi Certified Passpoint, is an approach to public access Wi-Fi by the Wi-Fi Alliance. The idea is for mobile devices to automatically join a Wi-Fi subscriber service whenever the user enters a Hotspot 2.0 area, in order to provide better bandwidth and services-on-demand to end-users and relieve carrier infrastructure of some traffic.

Hotspot 2.0 is based on the IEEE 802.11u standard, which is a set of protocols published in 2011 to enable cellular-like roaming. If the device supports 802.11u and is subscribed to a Hotspot 2.0 service it will automatically connect and roam.

Supported Devices

- Some Chinese tablet computers.
- Some THL smartphones.
- Apple mobile devices running iOS 7 and up.
- Some Samsung Galaxy smartphones.
- Windows 10 devices have full support for network discovery and connection.
- Windows 8 and Windows 8.1 lack network discovery, but supports connecting to a network when the credentials are known.

Billing

EDCF User-Priority-List

		Net traffic					
		low			high		
		Audio	Video	Data	Audio	Video	Data
User needs	time-critical	7	5	0	6	4	0
	not time-critical	-	-	2	-	-	2

The so-called "User-Fairness-Model " is a dynamic billing model, which allows a volume-based billing, charged only by the amount of payload (data, video, audio). Moreover, the tariff is classified by net traffic and user needs (Pommer, p. 116ff).

If the net traffic increases, then the user has to pay the next higher tariff class. By the way the user is asked for if he still wishes the session also by a higher traffic class.[dubious – discuss] Moreover, in time-critical applications (video, audio) a higher class fare is charged, than for non time-critical applications (such as reading Web pages, e-mail).

Tariff classes of the User-Fairness-Model

		Net traffic	
		low	high
User needs	time-critical	standard	exclusive
	not time-critical	low priced	standard

The "User-fairness model" can be implemented with the help of EDCF (IEEE 802.11e). A EDCF user priority list shares the traffic in 3 access categories (data, video, audio) and user priorities (UP) (Pommer, p. 117):

- Data [UP 0|2].
- Video [UP 5|4].
- Audio [UP 7|6].

Service-oriented provisioning for viable implementations.

Security Concerns

Some hotspots authenticate users; however, this does not prevent users from viewing network traffic using packet sniffers.

Some vendors provide a download option that deploys WPA support. This conflicts with enterprise configurations that have solutions specific to their internal WLAN.

In order to provide robust security to hotspot users, the Wi-Fi Alliance is developing a new hotspot program that aims to encrypt hotspot traffic with WPA2 security. The program was scheduled to launch in the first half of 2012.

Legal Concerns

Depending upon the location, providers of public hotspot access may have legal obligations, related to privacy requirements and liability for use for unlawful purposes. In countries where the internet is regulated or freedom of speech more restricted, there may be requirements such as licensing, logging, or recording of user information. Concerns may also relate to child safety, and social issues such as exposure to objectionable content, protection against cyberbullying and illegal behaviours, and prevention of perpetration of such behaviors by hotspot users themselves.

European Union

- Data Retention Directive Hotspot owners must retain key user statistics for 12 months.

- Directive on Privacy and Electronic Communications.

United Kingdom

- Data Protection Act 1998 The hotspot owner must retain individual's information within the confines of the law.

- Digital Economy Act 2010 Deals with, among other things, copyright infringement, and imposes fines of up to £250,000 for contravention.

Li-Fi

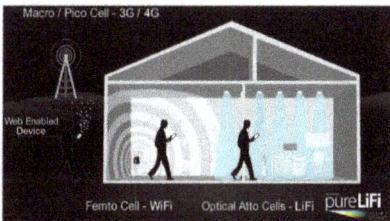

LiFi works in complement with existing and emerging wireless systems.

Light Fidelity (Li-Fi) is a bidirectional, high-speed and fully networked wireless communication technology similar to Wi-Fi. The term was coined by Harald Haas and is a form of visible light communication and a subset of optical wireless communications (OWC) and could be a complement to RF communication (Wi-Fi or cellular networks), or even a replacement in contexts of data broadcasting.

It is wireless and uses visible-light communication or infrared and near-ultraviolet instead of radio-frequency spectrum, part of optical wireless communications technology, which carries much more information, and has been proposed as a solution to the RF-bandwidth limitations.

Technology Details

This OWC technology uses light from light-emitting diodes (LEDs) as a medium to deliver networked, mobile, high-speed communication in a similar manner to Wi-Fi. The Li-Fi market is projected to have a compound annual growth rate of 82% from 2013 to 2018 and to be worth over $6 billion per year by 2018.

Visible light communications (VLC) works by switching the current to the LEDs off and on at a very high rate, too quick to be noticed by the human eye. Although Li-Fi LEDs would have to be kept on to transmit data, they could be dimmed to below human visibility while still emitting enough light to carry data. The light waves cannot penetrate walls which makes a much shorter range, though more secure from hacking, relative to Wi-Fi. Direct line of sight is not necessary for Li-Fi to transmit a signal; light reflected off the walls can achieve 70 Mbit/s.

Li-Fi has the advantage of being useful in electromagnetic sensitive areas such as in aircraft cabins, hospitals and nuclear power plants without causing electromagnetic interference. Both Wi-Fi and Li-Fi transmit data over the electromagnetic spectrum, but whereas Wi-Fi utilizes radio waves, Li-Fi uses visible light. While the US Federal Communications Commission has warned of a potential spectrum crisis because Wi-Fi is close to full capacity, Li-Fi has almost no limitations on capacity. The visible light spectrum is 10,000 times larger than the entire radio frequency spectrum. Researchers have reached data rates of over 10 Gbit/s, which is much faster than typical fast broadband in 2013. Li-Fi is expected to be ten times cheaper than Wi-Fi. Short range, low reliability and high installation costs are the potential downsides.

PureLiFi demonstrated the first commercially available Li-Fi system, the Li-1st, at the 2014 Mobile World Congress in Barcelona.

Bg-Fi is a Li-Fi system consisting of an application for a mobile device, and a simple consumer product, like an IoT (Internet of Things) device, with color sensor, microcontroller, and embedded software. Light from the mobile device display communicates to the color sensor on the consumer product, which converts the light into digital information. Light emitting diodes enable the consumer product to communicate synchronously with the mobile device.

History

Harald Haas, who teaches at the University of Edinburgh in Scotland, coined the term "Li-Fi" at his TED Global Talk where he introduced the idea of "Wireless data from every light". He is Chairman of Mobile Communications at the University of Edinburgh and co-founder of pureLiFi.

The general term visible light communication (VLC), whose history dates back to the 1880s, includes any use of the visible light portion of the electromagnetic spectrum to transmit information. The D-Light project at Edinburgh's Institute for Digital Communications was funded from January 2010 to January 2012. Haas promoted this technology in his 2011 TED Global talk and helped start a company to market it. PureLiFi, formerly pureVLC, is an original equipment manufacturer (OEM) firm set up to commercialize Li-Fi products for integration with existing LED-lighting systems.

In October 2011, companies and industry groups formed the Li-Fi Consortium, to promote high-speed optical wireless systems and to overcome the limited amount of radio-based wireless spectrum available by exploiting a completely different part of the electromagnetic spectrum.

A number of companies offer uni-directional VLC products, which is not the same as Li-Fi - a term defined by the IEEE 802.15.7r1 standardization committee.

VLC technology was exhibited in 2012 using Li-Fi. By August 2013, data rates of over 1.6 Gbit/s were demonstrated over a single color LED. In September 2013, a press release said that Li-Fi, or VLC systems in general, do not require line-of-sight conditions. In October 2013, it was reported Chinese manufacturers were working on Li-Fi development kits.

In April 2014, the Russian company Stins Coman announced the development of a Li-Fi wireless local network called BeamCaster. Their current module transfers data at 1.25 gigabytes per second but they fore boosting speeds up to 5 GB/second in the near future. In 2014 a new record was established by Sisoft (a Mexican company) that was able to transfer data at speeds of up to 10 Gbit/s across a light spectrum emitted by LED lamps.

Standards

Like Wi-Fi, Li-Fi is wireless and uses similar 802.11 protocols; but it uses visible light communication (instead of radio frequency waves), which has much wider bandwidth.

One part of VLC is modeled after communication protocols established by the IEEE 802 workgroup. However, the IEEE 802.15.7 standard is out-of-date, it fails to consider the latest technological developments in the field of optical wireless communications, specifically with the introduction of optical orthogonal frequency-division multiplexing (O-OFDM) modulation methods which have been optimized for data rates, multiple-access and energy efficiency. The introduction of O-OFDM means that a new drive for standardization of optical wireless communications is required.

Nonetheless, the IEEE 802.15.7 standard defines the physical layer (PHY) and media access control (MAC) layer. The standard is able to deliver enough data rates to transmit audio, video and multimedia services. It takes into account optical transmission mobility, its compatibility with artificial lighting present in infrastructures, and the interference which may be generated by ambient lighting. The MAC layer permits using the link with the other layers as with the TCP/IP protocol.

The standard defines three PHY layers with different rates:

- The PHY I was established for outdoor application and works from 11.67 kbit/s to 267.6 kbit/s.

- The PHY II layer permits reaching data rates from 1.25 Mbit/s to 96 Mbit/s.

- The PHY III is used for many emissions sources with a particular modulation method called color shift keying (CSK). PHY III can deliver rates from 12 Mbit/s to 96 Mbit/s.

The modulation formats recognized for PHY I and PHY II are on-off keying (OOK) and variable pulse position modulation (VPPM). The Manchester coding used for the PHY I and PHY II layers includes the clock inside the transmitted data by representing a logic 0 with an OOK symbol "01" and a logic 1 with an OOK symbol "10", all with a DC component. The DC component avoids light extinction in case of an extended run of logic 0's.

The first VLC smartphone prototype was presented at the Consumer Electronics Show in Las Vegas from January 7–10 in 2014. The phone uses SunPartner's Wysips CONNECT, a technique that converts light waves into usable energy, making the phone capable of receiving and decoding signals without drawing on its battery. A clear thin layer of crystal glass can be added to small screens like watches and smartphones that make them solar powered. Smartphones could gain 15% more battery life during a typical day. This first smartphones using this technology should arrive in 2015. This screen can also receive VLC signals as well as the smartphone camera. The cost of these screens per smartphone is between $2 and $3, much cheaper than most new technology.

Philips lighting company has developed a VLC system for shoppers at stores. They have to download an app on their smartphone and then their smartphone works with the LEDs in the store. The

LEDs can pinpoint where they are located in the store and give them corresponding coupons and information based on which aisle they are on and what they are looking at.

Bluetooth

Bluetooth is a wireless technology standard for exchanging data over short distances (using short-wavelength UHF radio waves in the ISM band from 2.4 to 2.485 GHz) from fixed and mobile devices, and building personal area networks (PANs). Invented by telecom vendor Ericsson in 1994, it was originally conceived as a wireless alternative to RS-232 data cables. It can connect several devices, overcoming problems of synchronization.

Bluetooth is managed by the Bluetooth Special Interest Group (SIG), which has more than 25,000 member companies in the areas of telecommunication, computing, networking, and consumer electronics. The IEEE standardized Bluetooth as IEEE 802.15.1, but no longer maintains the standard. The Bluetooth SIG overs development of the specification, manages the qualification program, and protects the trademarks. A manufacturer must make a device meet Bluetooth SIG standards to market it as a Bluetooth device. A network of patents apply to the technology, which are licensed to individual qualifying devices.

Origin

The development of the "short-link" radio technology, later named Bluetooth, was initiated in 1989 by Dr. Nils Rydbeck, CTO at Ericsson Mobile in Lund, and Dr. Johan Ullman. The purpose was to develop wireless headsets, according to two inventions by Johan Ullman, SE 8902098-6, issued 1989-06-12 and SE 9202239, issued 1992-07-24. Nils Rydbeck tasked Tord Wingren with specifying and Jaap Haartsen and Sven Mattisson with developing. Both were working for Ericsson in Lund, Sweden. The specification is based on frequency-hopping spread spectrum technology.

Name and Logo

The name "Bluetooth" is an Anglicised version of the Scandinavian *Blåtand/Blåtann* (Old Norse *blátⓩnn*), the epithet of the tenth-century king Harald Bluetooth who united dissonant Danish tribes into a single kingdom and, according to legend, introduced Christianity as well. The idea of this name was proposed in 1997 by Jim Kardach who developed a system that would allow mobile phones to communicate with computers. At the time of this proposal he was reading Frans G. Bengtsson's historical novel *The Long Ships* about Vikings and King Harald Bluetooth. The implication is that Bluetooth does the same with communications protocols, uniting them into one universal standard.

The Bluetooth logo is a bind rune merging the Younger Futhark runes ᚼ (Hagall) and ᛒ (Bjarkan), Harald's initials.

Implementation

Bluetooth operates at frequencies between 2402 and 2480 MHz, or 2400 and 2483.5 MHz including guard bands 2 MHz wide at the bottom end and 3.5 MHz wide at the top. This is in the globally unlicensed (but not unregulated) Industrial, Scientific and Medical (ISM) 2.4 GHz short-range radio frequency band. Bluetooth uses a radio technology called frequency-hopping spread spectrum. Bluetooth divides transmitted data into packets, and transmits each packet on one of 79 designated Bluetooth channels. Each channel has a bandwidth of 1 MHz. It usually performs 800 hops per second, with Adaptive Frequency-Hopping (AFH) enabled. Bluetooth low energy uses 2 MHz spacing, which accommodates 40 channels.

Originally, Gaussian frequency-shift keying (GFSK) modulation was the only modulation scheme available. Since the introduction of Bluetooth 2.0+EDR, π/4-DQPSK (Differential Quadrature Phase Shift Keying) and 8DPSK modulation may also be used between compatible devices. Devices functioning with GFSK are said to be operating in basic rate (BR) mode where an instantaneous data rate of 1 Mbit/s is possible. The term Enhanced Data Rate (EDR) is used to describe π/4-DPSK and 8DPSK schemes, each giving 2 and 3 Mbit/s respectively. The combination of these (BR and EDR) modes in Bluetooth radio technology is classified as a "BR/EDR radio".

Bluetooth is a packet-based protocol with a master-slave structure. One master may communicate with up to seven slaves in a piconet. All devices share the master's clock. Packet exchange is based on the basic clock, defined by the master, which ticks at 312.5 μs intervals. Two clock ticks make up a slot of 625 μs, and two slots make up a slot pair of 1250 μs. In the simple case of single-slot packets the master transmits in even slots and receives in odd slots. The slave, conversely, receives in even slots and transmits in odd slots. Packets may be 1, 3 or 5 slots long, but in all cases the master's transmission begins in even slots and the slave's in odd slots.

The above is valid for "classic" BT. Bluetooth Low Energy, introduced in the 4.0 specification, uses the same spectrum but somewhat differently; Bluetooth low energy#Radio interface.

Communication and Connection

A master Bluetooth device can communicate with a maximum of seven devices in a piconet (an ad-hoc computer network using Bluetooth technology), though not all devices reach this maximum. The devices can switch roles, by agreement, and the slave can become the master (for example, a headset initiating a connection to a phone necessarily begins as master—as initiator of the connection—but may subsequently operate as slave).

The Bluetooth Core Specification provides for the connection of two or more piconets to form a scatternet, in which certain devices simultaneously play the master role in one piconet and the slave role in another.

At any given time, data can be transferred between the master and one other device (except for the little-used broadcast mode.) The master chooses which slave device to address; typically, it switches rapidly from one device to another in a round-robin fashion. Since it is the master that chooses which slave to address, whereas a slave is (in theory) supposed to listen in each receive slot, being a master is a lighter burden than being a slave. Being a master of seven slaves is possible; being a slave of more than one master is difficult. The specification is vague as to required behavior in scatternets.

Uses

Class	Max. permitted power		Typ. range (m)
	(mW)	(dBm)	
1	100	20	~100
2	2.5	4	~10
3	1	0	~1
4	0.5	-3	~0.5

Bluetooth is a standard wire-replacement communications protocol primarily designed for low-power consumption, with a short range based on low-cost transceiver microchips in each device. Because the devices use a radio (broadcast) communications system, they do not have to be in visual line of sight of each other, however a *quasi optical* wireless path must be viable. Range is power-class-dependent, but effective ranges vary in practice; the table on the right.

Officially Class 3 radios have a range of up to 1 metre (3 ft), Class 2, most commonly found in mobile devices, 10 metres (33 ft), and Class 1, primarily for industrial use cases,100 metres (300 ft). Bluetooth Marketing qualifies that Class 1 range is in most cases 20–30 metres (66–98 ft), and Class 2 range 5–10 metres (16–33 ft).

Bluetooth Version	Maximum Speed	Maximum Range
3.0	25 Mbit/s	
4.0	25 Mbit/s	200 feet(60.96 m)
5.0	50 Mbit/s	800 feet(243.84 m)

The effective range varies due to propagation conditions, material coverage, production sample variations, antenna configurations and battery conditions. Most Bluetooth applications are for indoor conditions, where attenuation of walls and signal fading due to signal reflections make the range far lower than specified line-of-sight ranges of the Bluetooth products. Most Bluetooth applications are battery powered Class 2 devices, with little difference in range whether the other end of the link is a Class 1 or Class 2 device as the lower powered device tends to set the range limit. In some cases the effective range of the data link can be extended when a Class 2 device is connecting to a Class 1 transceiver with both higher sensitivity and transmission power than a typical Class 2 device. Mostly, however, the Class 1 devices have a similar sensitivity to Class 2 devices. Connecting two Class 1 devices with both high sensitivity and high power can allow ranges far in excess of the typical 100m, depending on the throughput required by the application. Some such devices allow open field ranges of up to 1 km and beyond between two similar devices without exceeding legal emission limits.

The Bluetooth Core Specification mandates a range of not less than 10 metres (33 ft), but there is no upper limit on actual range. Manufacturers' implementations can be tuned to provide the range needed for each case.

Bluetooth Profiles

To use Bluetooth wireless technology, a device must be able to interpret certain Bluetooth profiles, which are definitions of possible applications and specify general behaviours that Bluetooth-enabled devices use to communicate with other Bluetooth devices. These profiles include settings to parametrize and to control the communication from start. Adherence to profiles saves the time for transmitting the parameters anew before the bi-directional link becomes effective. There are a wide range of Bluetooth profiles that describe many different types of applications or use cases for devices.

List of Applications

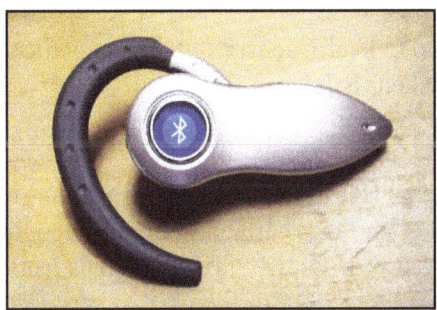

A typical Bluetooth mobile phone headset.

- Wireless control of and communication between a mobile phone and a handsfree headset. This was one of the earliest applications to become popular.

- Wireless control of and communication between a mobile phone and a Bluetooth compatible car stereo system.

- Wireless control of and communication with iOS and Android device phones, tablets and portable wireless speakers.

- Wireless Bluetooth headset and Intercom. Idiomatically, a headset is sometimes called "a Bluetooth".

- Wireless streaming of audio to headphones with or without communication capabilities.

- Wireless streaming of data collected by Bluetooth-enabled fitness devices to phone or PC.

- Wireless networking between PCs in a confined space and where little bandwidth is required.

- Wireless communication with PC input and output devices, the most common being the mouse, keyboard and printer.

- Transfer of files, contact details, calendar appointments, and reminders between devices with OBEX.

- Replacement of previous wired RS-232 serial communications in test equipment, GPS receivers, medical equipment, bar code scanners, and traffic control devices.

- For controls where infrared was often used.

- For low bandwidth applications where higher USB bandwidth is not required and cable-free connection desired.

- Sending small advertisements from Bluetooth-enabled advertising hoardings to other, discoverable, Bluetooth devices.

- Wireless bridge between two Industrial Ethernet (*e.g.*, PROFINET) networks.

- Seventh and eighth generation game consoles such as Nintendo's Wii, and Sony's PlayStation 3 use Bluetooth for their respective wireless controllers.

- Dial-up internet access on personal computers or PDAs using a data-capable mobile phone as a wireless modem.

- Short range transmission of health sensor data from medical devices to mobile phone, set-top box or dedicated telehealth devices.

- Allowing a DECT phone to ring and answer calls on behalf of a nearby mobile phone.

- Real-time location systems (RTLS), are used to track and identify the location of objects in real-time using "Nodes" or "tags" attached to, or embedded in the objects tracked, and "Readers" that receive and process the wireless signals from these tags to determine their locations.

- Personal security application on mobile phones for prevention of theft or loss of items. The protected item has a Bluetooth marker (*e.g.*, a tag) that is in constant communication with the phone. If the connection is broken (the marker is out of range of the phone) then an alarm is raised. This can also be used as a man overboard alarm. A product using this technology has been available since 2009.

- Calgary, Alberta, Canada's Roads Traffic division uses data collected from travelers' Bluetooth devices to predict travel times and road congestion for motorists.

- Wireless transmission of audio (a more reliable alternative to FM transmitters).

Bluetooth vs. Wi-Fi (IEEE 802.11)

Bluetooth and Wi-Fi (the brand name for products using IEEE 802.11 standards) have some similar applications: setting up networks, printing, or transferring files. Wi-Fi is intended as a replacement for high speed cabling for general local area network access in work areas or home. This category of applications is sometimes called wireless local area networks (WLAN). Bluetooth was intended for portable equipment and its applications. The category of applications is outlined as the wireless personal area network (WPAN). Bluetooth is a replacement for cabling in a variety of personally carried applications in any setting, and also works for fixed location applications such as smart energy functionality in the home (thermostats, etc.).

Wi-Fi and Bluetooth are to some extent complementary in their applications and usage. Wi-Fi is usually access point-centered, with an asymmetrical client-server connection with all traffic routed through the access point, while Bluetooth is usually symmetrical, between two Bluetooth devices. Bluetooth serves well in simple applications where two devices need to connect with minimal configuration like a button press, as in headsets and remote controls, while Wi-Fi suits better in applications where some degree of client configuration is possible and high speeds are required, especially for network access through an access node. However, Bluetooth access points do exist and ad-hoc connections are possible with Wi-Fi though not as simply as with Bluetooth. Wi-Fi Direct was recently developed to add a more Bluetooth-like ad-hoc functionality to Wi-Fi.

Devices

A Bluetooth USB dongle with a 100 m range.

Bluetooth exists in many products, such as telephones, tablets, media players, robotics systems, handheld, laptops and console gaming equipment, and some high definition headsets, modems, and watches. The technology is useful when transferring information between two or more devices that are near each other in low-bandwidth situations. Bluetooth is commonly used to transfer sound data with telephones (i.e., with a Bluetooth headset) or byte data with hand-held computers (transferring files).

Bluetooth protocols simplify the discovery and setup of services between devices. Bluetooth devices can advertise all of the services they provide. This makes using services easier, because more of the security, network address and permission configuration can be automated than with many other network types.

Computer Requirements

A typical Bluetooth USB dongle.

A personal computer that does not have embedded Bluetooth can use a Bluetooth adapter that enables the PC to communicate with Bluetooth devices. While some desktop computers and most recent laptops come with a built-in Bluetooth radio, others require an external adapter, typically in the form of a small USB "dongle."

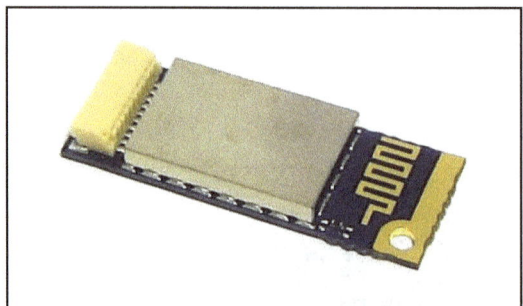

An internal notebook Bluetooth card (14×36×4 mm).

Unlike its predecessor, IrDA, which requires a separate adapter for each device, Bluetooth lets multiple devices communicate with a computer over a single adapter.

Operating System Implementation

For Microsoft platforms, Windows XP Service Pack 2 and SP3 releases work natively with Bluetooth v1.1, v2.0 and v2.0+EDR. Previous versions required users to install their Bluetooth adapter's own drivers, which were not directly supported by Microsoft. Microsoft's own Bluetooth dongles (packaged with their Bluetooth computer devices) have no external drivers and thus require at least Windows XP Service Pack 2. Windows Vista RTM/SP1 with the Feature Pack for Wireless or Windows Vista SP2 work with Bluetooth v2.1+EDR. Windows 7 works with Bluetooth v2.1+EDR and Extended Inquiry Response (EIR).

The Windows XP and Windows Vista/Windows 7 Bluetooth stacks support the following Bluetooth profiles natively: PAN, SPP, DUN, HID, HCRP. The Windows XP stack can be replaced by a third party stack that supports more profiles or newer Bluetooth versions. The Windows Vista/Windows 7 Bluetooth stack supports vendor-supplied additional profiles without requiring that the Microsoft stack be replaced.

Apple products have worked with Bluetooth since Mac OS X v10.2, which was released in 2002.

Linux has two popular Bluetooth stacks, BlueZ and Affix. The BlueZ stack is included with most Linux kernels and was originally developed by Qualcomm. The Affix stack was developed by Nokia.

FreeBSD features Bluetooth since its v5.0 release.

NetBSD features Bluetooth since its v4.0 release. Its Bluetooth stack has been ported to OpenBSD as well.

Specifications and Features

The specifications were formalized by the Bluetooth Special Interest Group (SIG). The SIG was formally announced on 20 May 1998. Today it has a membership of over 30,000 companies worldwide. It was established by Ericsson, IBM, Intel, Toshiba and Nokia, and later joined by many other companies.

All versions of the Bluetooth standards support downward compatibility. That lets the latest standard cover all older versions.

The Bluetooth Core Specification Working Group (CSWG) produces mainly 4 kinds of specifications:

- The Bluetooth Core Specification, release cycle is typically a few years in between.
- Core Specification Addendum (CSA), release cycle can be as tight as a few times per year.
- Core Specification Supplements (CSS), can be released very quickly.
- Errata.

Bluetooth v1.0 and v1.0B

Versions 1.0 and 1.0B had many problems and manufacturers had difficulty making their products interoperable. Versions 1.0 and 1.0B also included mandatory Bluetooth hardware device address (BD_ADDR) transmission in the Connecting process (rendering anonymity impossible at the protocol level), which was a major setback for certain services planned for use in Bluetooth environments.

Bluetooth v1.1

- Ratified as IEEE Standard 802.15.1–2002.
- Many errors found in the v1.0B specifications were fixed.
- Added possibility of non-encrypted channels.
- Received Signal Strength Indicator (RSSI).

Bluetooth v1.2

Major enhancements include the following:

- Faster Connection and Discovery.
- *Adaptive frequency-hopping spread spectrum (AFH)*, which improves resistance to radio frequency interference by avoiding the use of crowded frequencies in the hopping sequence.
- Higher transmission speeds in practice, up to 721 kbit/s, than in v1.1.
- Extended Synchronous Connections (eSCO), which improve voice quality of audio links by allowing retransmissions of corrupted packets, and may optionally increase audio latency to provide better concurrent data transfer.
- Host Controller Interface (HCI) operation with three-wire UART.
- Ratified as IEEE Standard 802.15.1–2005.
- Introduced Flow Control and Retransmission Modes for L2CAP.

Bluetooth v2.0 + EDR

This version of the Bluetooth Core Specification was released in 2004. The main difference is the introduction of an Enhanced Data Rate (EDR) for faster data transfer. The nominal rate of EDR is about 3 Mbit/s, although the practical data transfer rate is 2.1 Mbit/s. EDR uses a combination of GFSK and Phase Shift Keying modulation (PSK) with two variants, $\pi/4$-DQPSK and 8DPSK. EDR can provide a lower power consumption through a reduced duty cycle.

The specification is published as *Bluetooth v2.0 + EDR*, which implies that EDR is an optional feature. Aside from EDR, the v2.0 specification contains other minor improvements, and products may claim compliance to "Bluetooth v2.0" without supporting the higher data rate. At least one commercial device states "Bluetooth v2.0 without EDR" on its data sheet.

Bluetooth v2.1 + EDR

Bluetooth Core Specification Version 2.1 + EDR was adopted by the Bluetooth SIG on 26 July 2007.

The headline feature of v2.1 is secure simple pairing (SSP): this improves the pairing experience for Bluetooth devices, while increasing the use and strength of security. the section on Pairing below for more details.

Version 2.1 allows various other improvements, including "Extended inquiry response" (EIR), which provides more information during the inquiry procedure to allow better filtering of devices before connection; and sniff subrating, which reduces the power consumption in low-power mode.

Bluetooth v3.0 + HS

Version 3.0 + HS of the Bluetooth Core Specification was adopted by the Bluetooth SIG on 21 April 2009. Bluetooth v3.0 + HS provides theoretical data transfer speeds of up to 24 Mbit/s, though not over the Bluetooth link itself. Instead, the Bluetooth link is used for negotiation and establishment, and the high data rate traffic is carried over a colocated 802.11 link.

The main new feature is AMP (Alternative MAC/PHY), the addition of 802.11 as a high speed transport. The High-Speed part of the specification is not mandatory, and hence only devices that display the "+HS" logo actually support Bluetooth over 802.11 high-speed data transfer. A Bluetooth v3.0 device without the "+HS" suffix is only required to support features introduced in Core Specification Version 3.0 or earlier Core Specification Addendum 1.

L2CAP **Enhanced Modes**

Enhanced Retransmission Mode (ERTM) implements reliable L2CAP channel, while Streaming Mode (SM) implements unreliable channel with no retransmission or flow control. Introduced in Core Specification Addendum 1.

Alternative MAC/PHY

Enables the use of alternative MAC and PHYs for transporting Bluetooth profile data. The Bluetooth radio is still used for device discovery, initial connection and profile configuration. However, when large quantities of data must be sent, the high speed alternative MAC PHY 802.11 (typically associated with Wi-Fi) transports the data. This means that Bluetooth uses proven low power connection models when the system is idle, and the faster radio when it must send large quantities of data. AMP links require enhanced L2CAP modes.

Unicast Connectionless Data

Permits sending service data without establishing an explicit L2CAP channel. It is intended for use by applications that require low latency between user action and reconnection/transmission of data. This is only appropriate for small amounts of data.

Enhanced Power Control

Updates the power control feature to remove the open loop power control, and also to clarify ambiguities in power control introduced by the new modulation schemes added for EDR. Enhanced power control removes the ambiguities by specifying the behaviour that is expected. The feature also adds closed loop power control, meaning RSSI filtering can start as the response is received. Additionally, a "go straight to maximum power" request has been introduced. This is expected to deal with the headset link loss issue typically observed when a user puts their phone into a pocket on the opposite side to the headset.

Ultra-wideband

The high speed (AMP) feature of Bluetooth v3.0 was originally intended for UWB, but the WiMedia Alliance, the body responsible for the flavor of UWB intended for Bluetooth, announced in March 2009 that it was disbanding, and ultimately UWB was omitted from the Core v3.0 specification.

On 16 March 2009, the WiMedia Alliance announced it was entering into technology transfer agreements for the WiMedia Ultra-wideband (UWB) specifications. WiMedia has transferred all current and future specifications, including work on future high speed and power optimized implementations, to the Bluetooth Special Interest Group (SIG), Wireless USB Promoter Group and the USB Implementers Forum. After successful completion of the technology transfer, marketing, and related administrative items, the WiMedia Alliance ceased operations.

In October 2009 the Bluetooth Special Interest Group suspended development of UWB as part of the alternative MAC/PHY, Bluetooth v3.0 + HS solution. A small, but significant, number of former WiMedia members had not and would not sign up to the necessary agreements for the IP transfer. The Bluetooth SIG is now in the process of evaluating other options for its longer term roadmap.

Bluetooth v4.0

The Bluetooth SIG completed the Bluetooth Core Specification version 4.0 (called Bluetooth Smart) and has been adopted as of 30 June 2010. It includes *Classic Bluetooth, Bluetooth high speed* and *Bluetooth low energy* protocols. Bluetooth high speed is based on Wi-Fi, and Classic Bluetooth consists of legacy Bluetooth protocols.

Bluetooth low energy, previously known as Wibree, is a subset of Bluetooth v4.0 with an entirely new protocol stack for rapid build-up of simple links. As an alternative to the Bluetooth standard protocols that were introduced in Bluetooth v1.0 to v3.0, it is aimed at very low power applications running off a coin cell. Chip designs allow for two types of implementation, dual-mode, single-mode and enhanced past versions. The provisional names *Wibree* and *Bluetooth ULP* (Ultra Low Power) were abandoned and the BLE name was used for a while. In late 2011, new logos "Bluetooth Smart Ready" for hosts and "Bluetooth Smart" for sensors were introduced as the general-public face of BLE.

- In a single-mode implementation, only the low energy protocol stack is implemented. ST-Microelectronics, AMICCOM, CSR, Nordic Semiconductor and Texas Instruments have released single mode Bluetooth low energy solutions.

- In a dual-mode implementation, Bluetooth Smart functionality is integrated into an existing Classic Bluetooth controller. As of March 2011, the following semiconductor companies have announced the availability of chips meeting the standard: Qualcomm-Atheros, CSR, Broadcom and Texas Instruments. The compliant architecture shares all of Classic Bluetooth's existing radio and functionality resulting in a negligible cost increase compared to Classic Bluetooth.

Cost-reduced single-mode chips, which enable highly integrated and compact devices, feature a lightweight Link Layer providing ultra-low power idle mode operation, simple device discovery, and reliable point-to-multipoint data transfer with advanced power-save and secure encrypted connections at the lowest possible cost.

General improvements in version 4.0 include the changes necessary to facilitate BLE modes, as well the Generic Attribute Profile (GATT) and Security Manager (SM) services with AES Encryption.

Core Specification Addendum 2 was unveiled in December 2011; it contains improvements to the audio Host Controller Interface and to the High Speed (802.11) Protocol Adaptation Layer.

Core Specification Addendum 3 revision 2 has an adoption date of 24 July 2012.

Core Specification Addendum 4 has an adoption date of 12 February 2013.

Bluetooth v4.1

The Bluetooth SIG announced formal adoption of the Bluetooth v4.1 specification on 4 December 2013. This specification is an incremental software update to Bluetooth Specification v4.0, and not a hardware update. The update incorporates Bluetooth Core Specification Addenda (CSA 1, 2, 3 & 4) and adds new features that improve consumer usability. These include increased co-existence support for LTE, bulk data exchange rates—and aid developer innovation by allowing devices to support multiple roles simultaneously.

New features of this specification include:

- Mobile Wireless Service Coexistence Signaling.

- Train Nudging and Generalized Interlaced Scanning.

- Low Duty Cycle Directed Advertising.

- L2CAP Connection Oriented and Dedicated Channels with Credit Based Flow Control.

- Dual Mode and Topology.

- LE Link Layer Topology.

- 802.11n PAL.

- Audio Architecture Updates for Wide Band Speech.

- Fast Data Advertising Interval.

- Limited Discovery Time.

Notice that some features were already available in a Core Specification Addendum (CSA) before the release of v4.1.

Bluetooth v4.2

Bluetooth v4.2 was released on December 2, 2014. It Introduces some key features for IoT. Some features, such as Data Length Extension, require a hardware update. But some older Bluetooth hardware may receive some Bluetooth v4.2 features, such as privacy updates via firmware.

The major areas of improvement are:

- LE Data Packet Length Extension.

- LE Secure Connections.

- Link Layer Privacy.

- Link Layer Extended Scanner Filter Policies.

- IP connectivity for Bluetooth Smart devices to become available soon after the introduction of BT v4.2 via the new Internet Protocol Support Profile (IPSP).

- IPSP adds an IPv6 connection option for Bluetooth Smart, to support connected home and other IoT implementations.

Bluetooth v5

Bluetooth 5 was announced in June 2016. It will quadruple the range, double the speed, and provide an eight-fold increase in data broadcasting capacity of low energy Bluetooth connections, in addition to adding functionality for connectionless services like location-relevant information and navigation.

It is mainly focused on Internet of Things emerging technology. The release of products is scheduled for late 2016 to early 2017.

Technical Information

Bluetooth Protocol Stack

Bluetooth is defined as a layer protocol architecture consisting of core protocols, cable replacement protocols, telephony control protocols, and adopted protocols. Mandatory protocols for all

Bluetooth stacks are: LMP, L2CAP and SDP. In addition, devices that communicate with Bluetooth almost universally can use these protocols: HCI and RFCOMM.

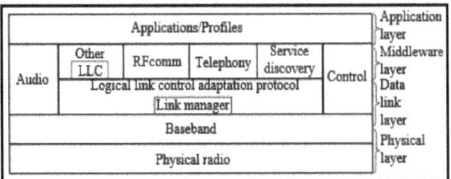

Bluetooth Protocol Stack.

LMP

The *Link Management Protocol* (LMP) is used for set-up and control of the radio link between two devices. Implemented on the controller.

L2CAP

The *Logical Link Control and Adaptation Protocol* (L2CAP) is used to multiplex multiple logical connections between two devices using different higher level protocols. Provides segmentation and reassembly of on-air packets.

In *Basic* mode, L2CAP provides packets with a payload configurable up to 64 kB, with 672 bytes as the default MTU, and 48 bytes as the minimum mandatory supported MTU.

In *Retransmission and Flow Control* modes, L2CAP can be configured either for isochronous data or reliable data per channel by performing retransmissions and CRC checks.

Bluetooth Core Specification Addendum 1 adds two additional L2CAP modes to the core specification. These modes effectively deprecate original Retransmission and Flow Control modes:

- Enhanced Retransmission Mode (ERTM): This mode is an improved version of the original retransmission mode. This mode provides a reliable L2CAP channel.

- Streaming Mode (SM): This is a very simple mode, with no retransmission or flow control. This mode provides an unreliable L2CAP channel.

Reliability in any of these modes is optionally and/or additionally guaranteed by the lower layer Bluetooth BDR/EDR air interface by configuring the number of retransmissions and flush timeout (time after which the radio flushes packets). In-order sequencing is guaranteed by the lower layer.

Only L2CAP channels configured in ERTM or SM may be operated over AMP logical links.

SDP

The *Service Discovery Protocol* (SDP) allows a device to discover services offered by other devices, and their associated parameters. For example, when you use a mobile phone with a Bluetooth headset, the phone uses SDP to determine which Bluetooth profiles the headset can use (Headset Profile, Hands Free Profile, Advanced Audio Distribution Profile (A2DP) etc.) and the protocol multiplexer settings needed for the phone to connect to the headset using each of them. Each service is identified by a Universally Unique Identifier (UUID), with official services (Bluetooth profiles) assigned a short form UUID (16 bits rather than the full 128).

RFCOMM

Radio Frequency Communications (RFCOMM) is a cable replacement protocol used to generate a virtual serial data stream. RFCOMM provides for binary data transport and emulates EIA-232 (formerly RS-232) control signals over the Bluetooth baseband layer, i.e. it is a serial port emulation.

RFCOMM provides a simple reliable data stream to the user, similar to TCP. It is used directly by many telephony related profiles as a carrier for AT commands, as well as being a transport layer for OBEX over Bluetooth.

Many Bluetooth applications use RFCOMM because of its widespread support and publicly available API on most operating systems. Additionally, applications that used a serial port to communicate can be quickly ported to use RFCOMM.

BNEP

The *Bluetooth Network Encapsulation Protocol* (BNEP) is used for transferring another protocol stack's data via an L2CAP channel. Its main purpose is the transmission of IP packets in the Personal Area Networking Profile. BNEP performs a similar function to SNAP in Wireless LAN.

AVCTP

The *Audio/Video Control Transport Protocol* (AVCTP) is used by the remote control profile to transfer AV/C commands over an L2CAP channel. The music control buttons on a stereo headset use this protocol to control the music player.

AVDTP

The *Audio/Video Distribution Transport Protocol* (AVDTP) is used by the advanced audio distribution profile to stream music to stereo headsets over an L2CAP channel intended for video distribution profile in the Bluetooth transmission.

TCS

The *Telephony Control Protocol – Binary* (TCS BIN) is the bit-oriented protocol that defines the call control signaling for the establishment of voice and data calls between Bluetooth devices. Additionally, "TCS BIN defines mobility management procedures for handling groups of Bluetooth TCS devices."

TCS-BIN is only used by the cordless telephony profile, which failed to attract implementers. As such it is only of historical interest.

Adopted Protocols

Adopted protocols are defined by other standards-making organizations and incorporated into

Bluetooth's protocol stack, allowing Bluetooth to code protocols only when necessary. The adopted protocols include:

- Point-to-Point Protocol (PPP): Internet standard protocol for transporting IP datagrams over a point-to-point link.

- TCP/IP/UDP: Foundation Protocols for TCP/IP protocol suite.

- Object Exchange Protocol (OBEX): Session-layer protocol for the exchange of objects, providing a model for object and operation representation.

- Wireless Application Environment/Wireless Application Protocol (WAE/WAP): WAE specifies an application framework for wireless devices and WAP is an open standard to provide mobile users access to telephony and information services.

Baseband Error Correction

Depending on packet type, individual packets may be protected by error correction, either 1/3 rate forward error correction (FEC) or 2/3 rate. In addition, packets with CRC will be retransmitted until acknowledged by automatic repeat request (ARQ).

Setting Up Connections

Any Bluetooth device in *discoverable mode* transmits the following information on demand:

- Device name.

- Device class.

- List of services.

- Technical information (for example: device features, manufacturer, Bluetooth specification used, clock offset).

Any device may perform an inquiry to find other devices to connect to, and any device can be configured to respond to such inquiries. However, if the device trying to connect knows the address of the device, it always responds to direct connection requests and transmits the information shown in the list above if requested. Use of a device's services may require pairing or acceptance by its owner, but the connection itself can be initiated by any device and held until it goes out of range. Some devices can be connected to only one device at a time, and connecting to them prevents them from connecting to other devices and appearing in inquiries until they disconnect from the other device.

Every device has a unique 48-bit address. However, these addresses are generally not shown in inquiries. Instead, friendly Bluetooth names are used, which can be set by the user. This name appears when another user scans for devices and in lists of paired devices.

Most cellular phones have the Bluetooth name set to the manufacturer and model of the phone by default. Most cellular phones and laptops show only the Bluetooth names and special programs are required to get additional information about remote devices. This can be confusing as, for example, there could be several cellular phones in range named T610.

Pairing and Bonding

Motivation

Many services offered over Bluetooth can expose private data or let a connecting party control the Bluetooth device. Security reasons make it necessary to recognize specific devices, and thus enable control over which devices can connect to a given Bluetooth device. At the same time, it is useful for Bluetooth devices to be able to establish a connection without user intervention (for example, as soon as in range).

To resolve this conflict, Bluetooth uses a process called *bonding*, and a bond is generated through a process called *pairing*. The pairing process is triggered either by a specific request from a user to generate a bond (for example, the user explicitly requests to "Add a Bluetooth device"), or it is triggered automatically when connecting to a service where (for the first time) the identity of a device is required for security purposes. These two cases are referred to as dedicated bonding and general bonding respectively.

Pairing often involves some level of user interaction. This user interaction confirms the identity of the devices. When pairing successfully completes, a bond forms between the two devices, enabling those two devices to connect to each other in the future without repeating the pairing process to confirm device identities. When desired, the user can remove the bonding relationship.

Implementation

During pairing, the two devices establish a relationship by creating a shared secret known as a *link key*. If both devices store the same link key, they are said to be *paired* or *bonded*. A device that wants to communicate only with a bonded device can cryptographically authenticate the identity of the other device, ensuring it is the same device it previously paired with. Once a link key is generated, an authenticated Asynchronous Connection-Less (ACL) link between the devices may be encrypted to protect exchanged data against eavesdropping. Users can delete link keys from either device, which removes the bond between the devices—so it is possible for one device to have a stored link key for a device it is no longer paired with.

Bluetooth services generally require either encryption or authentication and as such require pairing before they let a remote device connect. Some services, such as the Object Push Profile, elect not to explicitly require authentication or encryption so that pairing does not interfere with the user experience associated with the service use-cases.

Pairing Mechanisms

Pairing mechanisms changed significantly with the introduction of Secure Simple Pairing in Bluetooth v2.1. The following summarizes the pairing mechanisms:

- *Legacy pairing*: This is the only method available in Bluetooth v2.0 and before. Each device must enter a PIN code; pairing is only successful if both devices enter the same PIN code. Any 16-byte UTF-8 string may be used as a PIN code; however, not all devices may be capable of entering all possible PIN codes.

    ○ *Limited input devices*: The obvious example of this class of device is a Bluetooth

Hands-free headset, which generally have few inputs. These devices usually have a *fixed PIN*, for example "0000" or "1234", that are hard-coded into the device.

- *Numeric input devices*: Mobile phones are classic examples of these devices. They allow a user to enter a numeric value up to 16 digits in length.

- *Alpha-numeric input devices*: PCs and smartphones are examples of these devices. They allow a user to enter full UTF-8 text as a PIN code. If pairing with a less capable device the user must be aware of the input limitations on the other device, there is no mechanism available for a capable device to determine how it should limit the available input a user may use.

- *Secure Simple Pairing* (SSP): This is required by Bluetooth v2.1, although a Bluetooth v2.1 device may only use legacy pairing to interoperate with a v2.0 or earlier device. Secure Simple Pairing uses a form of public key cryptography, and some types can help protect against man in the middle, or MITM attacks. SSP has the following authentication mechanisms:

  - *Just works*: As the name implies, this method just works, with no user interaction. However, a device may prompt the user to confirm the pairing process. This method is typically used by headsets with very limited IO capabilities, and is more secure than the fixed PIN mechanism this limited set of devices uses for legacy pairing. This method provides no man-in-the-middle (MITM) protection.

  - *Numeric comparison*: If both devices have a display, and at least one can accept a binary yes/no user input, they may use Numeric Comparison. This method displays a 6-digit numeric code on each device. The user should compare the numbers to ensure they are identical. If the comparison succeeds, the user(s) should confirm pairing on the device(s) that can accept an input. This method provides MITM protection, assuming the user confirms on both devices and actually performs the comparison properly.

  - *Passkey Entry*: This method may be used between a device with a display and a device with numeric keypad entry (such as a keyboard), or two devices with numeric keypad entry. In the first case, the display is used to show a 6-digit numeric code to the user, who then enters the code on the keypad. In the second case, the user of each device enters the same 6-digit number. Both of these cases provide MITM protection.

  - *Out of band* (OOB): This method uses an external means of communication, such as Near Field Communication (NFC) to exchange some information used in the pairing process. Pairing is completed using the Bluetooth radio, but requires information from the OOB mechanism. This provides only the level of MITM protection that is present in the OOB mechanism.

SSP is considered simple for the following reasons:

- In most cases, it does not require a user to generate a passkey.

- For use-cases not requiring MITM protection, user interaction can be eliminated.

- For *numeric comparison*, MITM protection can be achieved with a simple equality comparison by the user.

- Using OOB with NFC enables pairing when devices simply get close, rather than requiring a lengthy discovery process.

Security Concerns

Prior to Bluetooth v2.1, encryption is not required and can be turned off at any time. Moreover, the encryption key is only good for approximately 23.5 hours; using a single encryption key longer than this time allows simple XOR attacks to retrieve the encryption key.

- Turning off encryption is required for several normal operations, so it is problematic to detect if encryption is disabled for a valid reason or for a security attack.

- Bluetooth v2.1 addresses this in the following ways.

- Encryption is required for all non-SDP (Service Discovery Protocol) connections.

- A new Encryption Pause and Resume feature is used for all normal operations that require that encryption be disabled. This enables easy identification of normal operation from security attacks.

- The encryption key must be refreshed before it expires.

Link keys may be stored on the device file system, not on the Bluetooth chip itself. Many Bluetooth chip manufacturers let link keys be stored on the device—however, if the device is removable, this means that the link key moves with the device.

Security

Overview

Bluetooth implements confidentiality, authentication and key derivation with custom algorithms based on the SAFER+ block cipher. Bluetooth key generation is generally based on a Bluetooth PIN, which must be entered into both devices. This procedure might be modified if one of the devices has a fixed PIN (e.g., for headsets or similar devices with a restricted user interface). During pairing, an initialization key or master key is generated, using the E22 algorithm. The E0 stream cipher is used for encrypting packets, granting confidentiality, and is based on a shared cryptographic secret, namely a previously generated link key or master key. Those keys, used for subsequent encryption of data sent via the air interface, rely on the Bluetooth PIN, which has been entered into one or both devices.

An overview of Bluetooth vulnerabilities exploits was published in 2007 by Andreas Becker.

In September 2008, the National Institute of Standards and Technology (NIST) published a Guide to Bluetooth Security as a reference for organizations. It describes Bluetooth security capabilities and how to secure Bluetooth technologies effectively. While Bluetooth has its benefits, it is susceptible to denial-of-service attacks, eavesdropping, man-in-the-middle attacks, message modification, and resource misappropriation. Users and organizations must evaluate their acceptable level of risk and incorporate security into the lifecycle of Bluetooth devices. To help mitigate risks, included in the NIST document are security checklists with guidelines and recommendations for creating and maintaining secure Bluetooth piconets, headsets, and smart card readers.

Bluetooth v2.1 – finalized in 2007 with consumer devices first appearing in 2009 – makes significant changes to Bluetooth's security, including pairing. the pairing mechanisms section for more about these changes.

Bluejacking

Bluejacking is the sending of either a picture or a message from one user to an unsuspecting user through *Bluetooth* wireless technology. Common applications include short messages, *e.g.*, "You've just been bluejacked!". Bluejacking does not involve the removal or alteration of any data from the device. Bluejacking can also involve taking control of a mobile device wirelessly and phoning a premium rate line, owned by the bluejacker. Security advances have alleviated this issue.

History of Security Concerns

2001–2004

In 2001, Jakobsson and Wetzel from Bell Laboratories discovered flaws in the Bluetooth pairing protocol and also pointed to vulnerabilities in the encryption scheme. In 2003, Ben and Adam Laurie from A.L. Digital Ltd. discovered that serious flaws in some poor implementations of Bluetooth security may lead to disclosure of personal data. In a subsequent experiment, Martin Herfurt from the trifinite.group was able to do a field-trial at the CeBIT fairgrounds, showing the importance of the problem to the world. A new attack called BlueBug was used for this experiment. In 2004 the first purported virus using Bluetooth to spread itself among mobile phones appeared on the Symbian OS. The virus was first described by Kaspersky Lab and requires users to confirm the installation of unknown software before it can propagate. The virus was written as a proof-of-concept by a group of virus writers known as "29A" and sent to anti-virus groups. Thus, it should be regarded as a potential (but not real) security threat to Bluetooth technology or Symbian OS since the virus has never spread outside of this system. In August 2004, a world-record-setting experiment showed that the range of Class 2 Bluetooth radios could be extended to 1.78 km (1.11 mi) with directional antennas and signal amplifiers. This poses a potential security threat because it enables attackers to access vulnerable Bluetooth devices from a distance beyond expectation. The attacker must also be able to receive information from the victim to set up a connection. No attack can be made against a Bluetooth device unless the attacker knows its Bluetooth address and which channels to transmit on, although these can be deduced within a few minutes if the device is in use.

2005

In January 2005, a mobile malware worm known as *John Cena.* began targeting mobile phones using Symbian OS (Series 60 platform) using Bluetooth enabled devices to replicate itself and spread to other devices. The worm is self-installing and begins once the mobile user approves the transfer of the file (velasco.sis) from another device. Once installed, the worm begins looking for other Bluetooth enabled devices to infect. Additionally, the worm infects other .SIS files on the device, allowing replication to another device through use of removable media (Secure Digital, Compact Flash, etc.). The worm can render the mobile device unstable.

In April 2005, Cambridge University security researchers published results of their actual implementation of passive attacks against the PIN-based pairing between commercial Bluetooth devices. They confirmed that attacks are practicably fast, and the Bluetooth symmetric key establishment method is vulnerable. To rectify this vulnerability, they designed an implementation that showed that stronger, asymmetric key establishment is feasible for certain classes of devices, such as mobile phones.

In June 2005, Yaniv Shaked and Avishai Wool published a paper describing both passive and active methods for obtaining the PIN for a Bluetooth link. The passive attack allows a suitably equipped attacker to eavesdrop on communications and spoof, if the attacker was present at the time of initial pairing. The active method makes use of a specially constructed message that must be inserted at a specific point in the protocol, to make the master and slave repeat the pairing process. After that, the first method can be used to crack the PIN. This attack's major weakness is that it requires the user of the devices under attack to re-enter the PIN during the attack when the device prompts them to. Also, this active attack probably requires custom hardware, since most commercially available Bluetooth devices are not capable of the timing necessary.

In August 2005, police in Cambridgeshire, England, issued warnings about thieves using Bluetooth enabled phones to track other devices left in cars. Police are advising users to ensure that any mobile networking connections are de-activated if laptops and other devices are left in this way.

2006

In April 2006, researchers from Secure Network and F-Secure published a report that warns of the large number of devices left in a visible state, and issued statistics on the spread of various Bluetooth services and the ease of spread of an eventual Bluetooth worm.

2007

In October 2007, at the Luxemburgish Hack.lu Security Conference, Kevin Finistere and Thierry Zoller demonstrated and released a remote root shell via Bluetooth on Mac OS X v10.3.9 and v10.4. They also demonstrated the first Bluetooth PIN and Linkkeys cracker, which is based on the research of Wool and Shaked.

Mitigation

Options to mitigate against Bluetooth security attacks include:

- Enable Bluetooth only when required.
- Enable Bluetooth discovery only when necessary, and disable discovery when finished.
- Do not enter link keys or PINs when unexpectedly prompted to do so.
- Remove paired devices when not in use.
- Regularly update firmware on Bluetooth-enabled devices.

Health Concerns

Bluetooth uses the microwave radio frequency spectrum in the 2.402 GHz to 2.480 GHz range. Maximum power output from a Bluetooth radio is 100 mW for class 1, 2.5 mW for class 2, and 1 mW for class 3 devices. Even the maximum power output of class 1 is a lower level than the lowest powered mobile phones. UMTS & W-CDMA outputs 250 mW, GSM1800/1900 outputs 1000 mW, and GSM850/900 outputs 2000 mW.

Interference Caused by USB 3.0

USB 3.0 devices, ports and cables have been proven to interfere with Bluetooth devices due to the electronic noise they release falling over the same operating band as Bluetooth. The close proximity of Bluetooth and USB 3.0 devices can result in a drop in throughput or complete connection loss of the Bluetooth device/s connected to a computer.

Various strategies can be applied to resolve the problem, ranging from simple solutions such as increasing the distance of USB 3.0 devices from any Bluetooth devices or purchasing better shielded USB cables. Other solutions include applying additional shielding to the internal USB components of a computer.

Bluetooth Award Programs

The Bluetooth Innovation World Cup, a marketing initiative of the Bluetooth Special Interest Group (SIG), was an international competition that encouraged the development of innovations for applications leveraging Bluetooth technology in sports, fitness and health care products. The aim of the competition was to stimulate new markets.

The Bluetooth Innovation World Cup morphed into the Bluetooth Breakthrough Awards in 2013. The Breakthrough Awards Bluetooth program highlights the most innovative products and applications available today, prototypes coming soon, and student-led projects in the making.

References

- "American Airlines and MobileStar Network to Deliver Wireless Internet Connectivity to American's Passengers". PR Newswire. 11 May 2000. Retrieved 13 April 2013

- Sherman, Joshua (30 October 2013). "How LED Light Bulbs could replace Wi-Fi". Digital Trends. Retrieved 29 November 2015

- "Global Visible Light Communication (VLC)/Li-Fi Technology Market worth $6,138.02 Million by 2018". MarketsandMarkets. 10 January 2013. Retrieved 29 November 2015

- Coetzee, Jacques (13 January 2013). "LiFi beats Wi-Fi with 1Gb wireless speeds over pulsing LEDs". Gearburn. Retrieved 29 November 2015.

- Vincent, James (29 October 2013). "Li-Fi revolution: internet connections using light bulbs are 250 times". The Independent. Retrieved 29 November 2015

- G. Miao, J. Zander, K-W Sung, and B. Slimane, Fundamentals of Mobile Data Networks, Cambridge University Press, ISBN 1107143217, 2016

- Vega, Anna (14 July 2014). "Li-fi record data transmission of 10Gbps set using LED lights". Engineering and Technology Magazine. Retrieved 29 November 2015

- Van Camp, Jeffrey (19 January 2014). "Wysips Solar Charging Screen Could Eliminate Chargers and Wi-Fi". Digital Trends. Retrieved 29 November 2015

- "Towards Energy-Awareness in Managing Wireless LAN Applications". IEEE/IFIP NOMS 2012: IEEE/IFIP Network Operations and Management Symposium. Retrieved 2014-08-11

- "Application Level Energy and Performance Measurements in a Wireless LAN". The 2011 IEEE/ACM International Conference on Green Computing and Communications. Retrieved 2014-08-11

- Rigg, Jamie (January 11, 2014). "Smartphone concept incorporates LiFi sensor for receiving light-based data". Engadget. Retrieved January 16, 2014

- Scarfone, K. & Padgette, J. (September 2008). "Guide to Bluetooth Security" (PDF). National Institute of Standards and Technology. Retrieved 3 July 2013

- Vilorio, Dennis. "You're a what? Tower Climber" (PDF). Occupational Outlook Quarterly. Archived (PDF) from the original on February 3, 2013. Retrieved December 6, 2013

4

Diverse Forms of Multiplexing

Multiplexing is a popular networking technique in which multiple analog and digital signals are combined into one signal over a shared medium. This allows for a large number of signals to be transmitted over one medium. Time-division multiplexing, frequency-division multiplexing, space-division multiplexing, polarization-multiplexing, orbital angular momentum multiplexing, wavelength-division multiplexing, etc. are all types of multiplexing which are thoroughly covered in this chapter.

Multiplexing is a way of sending multiple signals or streams of information over a communications link at the same time in the form of a single, complex signal; the receiver recovers the separate signals, a process called *demultiplexing* (or *demuxing*).

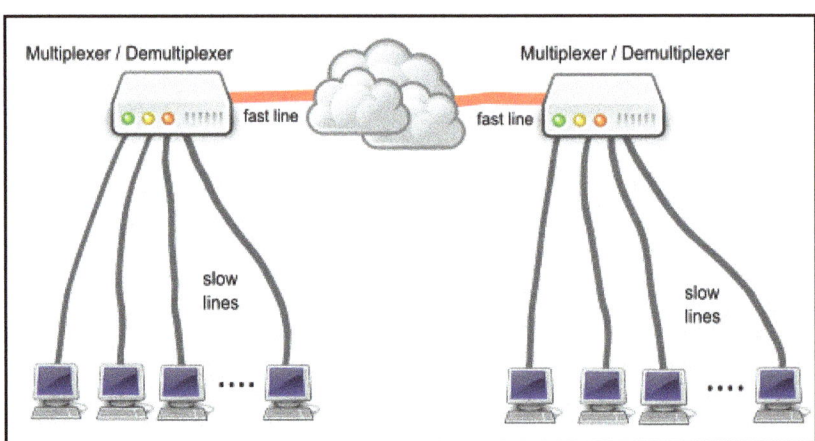

Networks use multiplexing for two reasons:

- To make it possible for any network device to talk to any other network device without having to dedicate a connection for each pair. This requires shared media;

- To make a scarce or expensive resource stretch further -- e.g., to send many signals down each cable or fiber strand running between major metropolitan areas, or across one satellite uplink.

In analog radio transmission, signals are commonly multiplexed using frequency-division multiplexing (FDM), in which the bandwidth on a communications link is divided into subchannels of different frequency widths, each carrying a signal at the same time in parallel. Analog cable TV works the same way, sending multiple channels of material down the same strands of coaxial cable.

Similarly, in some optical networks, data for different communications channels are sent on lightwaves of different wavelengths, a variety of multiplexing called *wave-length division multiplexing* (WDM).

The Multiplexer acts as a multiple-input and single-output switch. Multiple signals share one device or transmission conductor such as a copper wire or fiber optic cable. In telecommunications, the analog or digital signals transmitted on several communication channels by a multiplex method. These signals are single-output higher-speed signals. A 4-to-1 multiplexer contains four input signals and 2-to-1 multiplexer has two input signals and one output signal.

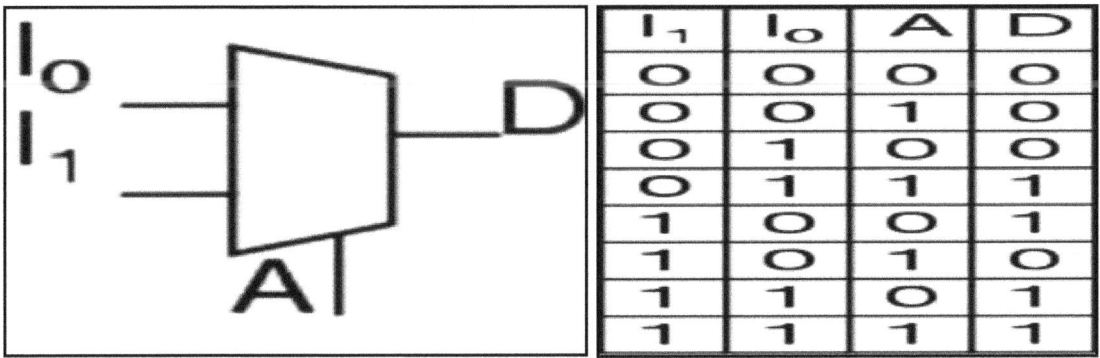

I_1	I_0	A	D
0	0	0	0
0	0	1	0
0	1	0	0
0	1	1	1
1	0	0	1
1	0	1	0
1	1	0	1
1	1	1	1

Schematic Symbol for Multiplexer.   Truth Table for 2 to 1 Multiplexer.

Multiplexers are also extended with same name conventions as DE multiplexers. A 4 to 1 multiplexer circuit is as below:

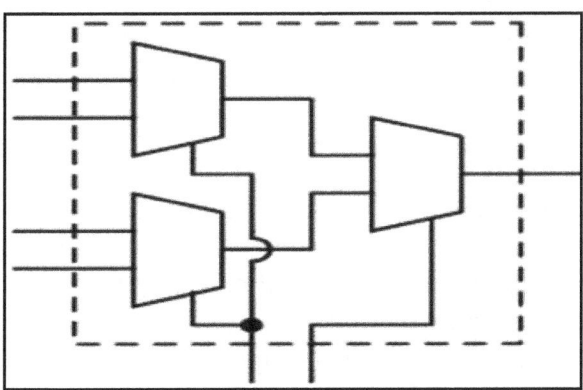

4 to 1 Multiplexer Circuit.

The different types of multiplexing technologies are as below:

- Wavelength Division Multiplexing (WDM).

- Frequency Division Multiplexing (FDM).

- Dense Wavelength Division Multiplexing (DWDM).

- Conventional Wavelength Division Multiplexing (CWDM).

- Reconfigurable Optical Add-Drop Multiplexer (ROADM).

- Orthogonal Frequency Division Multiplexing (OFDM).

- Add/Drop Multiplexing (ADM).

- Inverse Multiplexing (IMUX).

Types of Multiplexer

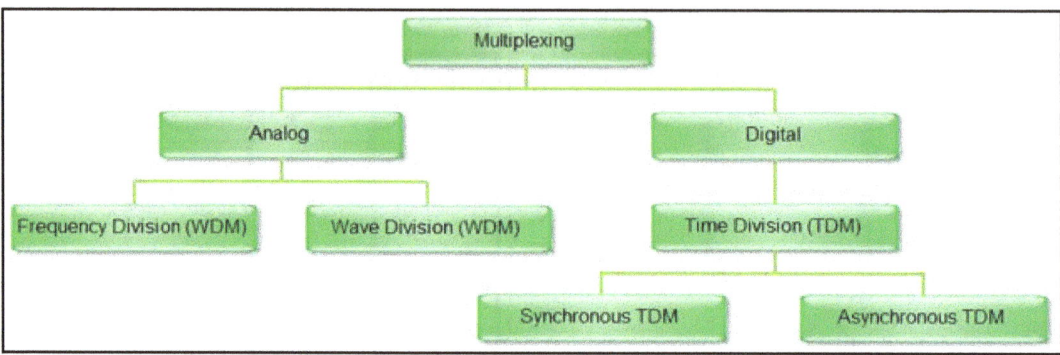

Frequency Division Multiplexing

Frequency Division Multiplexing is a technique which uses various frequencies to combine many streams of data for sending signals over a medium for communication purpose. It carries frequency to each data stream and later combines various modulated frequencies to transmission. Television Transmitters are the best example for FDM, which uses FDM to broad cast many channels at a time.

Wavelength Division Multiplexing

Wavelength Division Multiplexing (WDM) is analog multiplexing technique and it modulates many data streams on light spectrum. This multiplexing is used in optical fiber. It is FDM optical equivalent. Various signals in WDM are optical signal that will be light and were transmitted through optical fiber. WDM similar to FDM as it mixes many signals of different frequencies into single signal and transfer on one link. Wavelength of wave is reciprocal to its frequency, if wavelength increase then frequency decreases. Several light waves from many sources are united to get light signal which will be transmitted across channel to receiver.

The main principle in using prisms is that they bend a light beam depending on angle of incidence and frequency of light wave or ray. At receiver end the light signal is split into different light waves by demux. This type of merging and breaking of light wave made by a prism. Single prism is used at the end of sender for multiplexing and other prism is used at receiver end for demultiplexing as shown in figure.

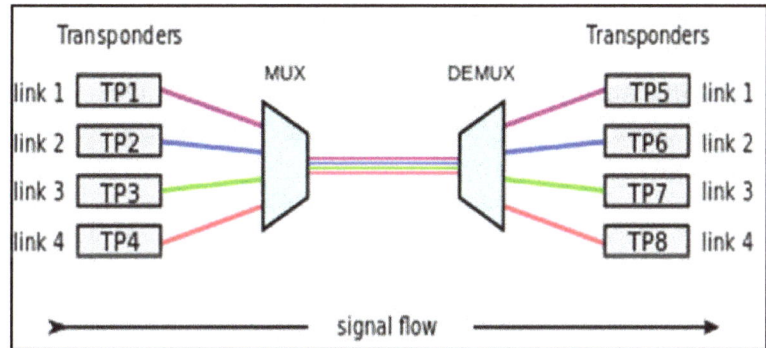
Wavelength Division Multiplexing.

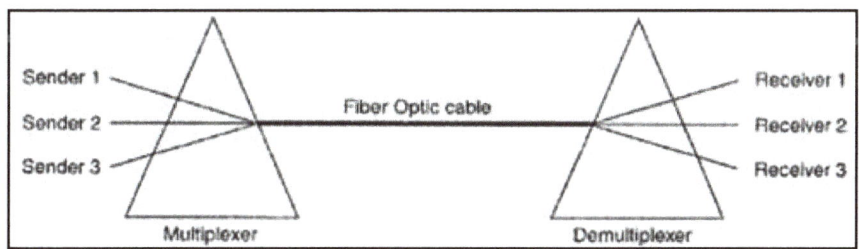
Usage of PRISM in WDM.

WDM used in Synchronous Optical Network (SONET). It utilizes various optical fiber lines that are multiplexed and demultiplexed.

Time Division Multiplexer

TDM is one of types of multiplexers which join data streams by allotting every stream different time slot in a set. It frequently transfers or sends various time slots in an order over one transmission channel. TDM attaches PCM data streams.

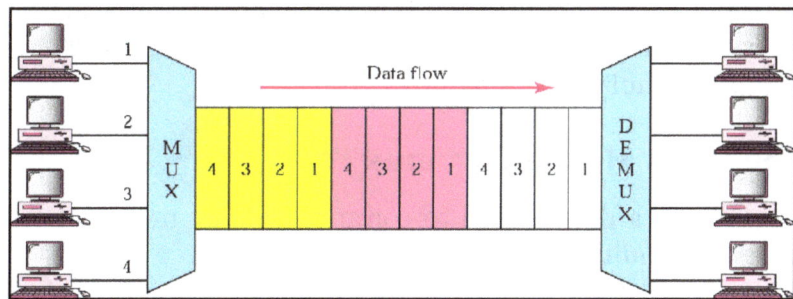
Time Division Multiplexer.

Dense Wavelength Division Multiplexer

In Dense Wavelength Division Multiplexing, an optical technology used to expand bandwidth onto fiber optic. Bit rate and protocol are independent and these are the main advantage of DWDM. Dense Wavelength Division Multiplexing (DWDM) operated by combining different signals simultaneously at different wavelengths. On fiber is changed to multiple fibers. By increasing the carrier capacity of fiber from 2.5Gb/s to 20 Gb/s, an eight OC 48 signals can be multiplexed into single fiber. Single fibers are able to transfer data at a speed upto 400 GB/s due to DWDM. DWDM transfers data or information in IP, SONET, ATM and Ethernet It also carries different type of traffic at a range of speeds on an optical channel.

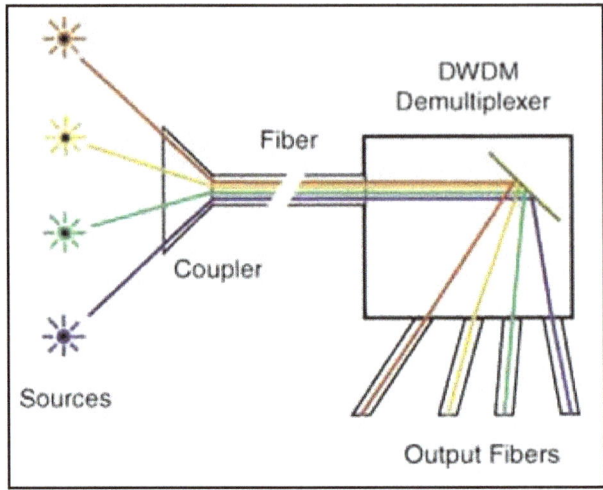

Dense Wavelength Division Multiplexer.

Statistical Multiplexer

It allows to share a single line of data for multiplexer RS-232 devices. Error correction will be performed in order to ensure the transmission an error-free one. The word "Statistical" refers to its capability to receive advantage of statics of many RS-232 devices means terminal and PC users. Each PC averages less than 5% of its potential data rate.

This type of multiplexer permits the sum of terminal and PC rates in which it extends composite link speed between multiplexers. This is due the reason that the keyboards are idle. These types of multiplexers requires buffer.

Difference between Mux and Demux

- A Multiplexer is a device used to communicate by means of a medium with combination of multiple signals.

- A DE multiplexer is a process of separating multiplexed signals from transmission line.

- These both Mux and DMux are mixed into single device which has the capability to process outgoing and incoming signals.

- An electronic multiplexer is a multiple-input, single-output switch.

- A DE multiplexer as a single-input, multiple-output switch.

- MUX allows many signals to share one device.

- Example: one A/D converter or one communication line.

Applications of Multiplexers

A Multiplexer is used in numerous applications like, where multiple data can be transmitted using a single line.

Communication System – A Multiplexer is used in communication systems, which has a transmission system and also a communication network. A Multiplexer is used to increase the efficiency of the communication system by allowing the transmission of data such as audio & video data from different channels via cables and single lines.

Computer Memory – A Multiplexer is used in computer memory to keep up a vast amount of memory in the computers, and also to decrease the number of copper lines necessary to connect the memory to other parts of the computer.

Telephone Network – A multiplexer is used in telephone networks to integrate the multiple audio signals on a single line of transmission.

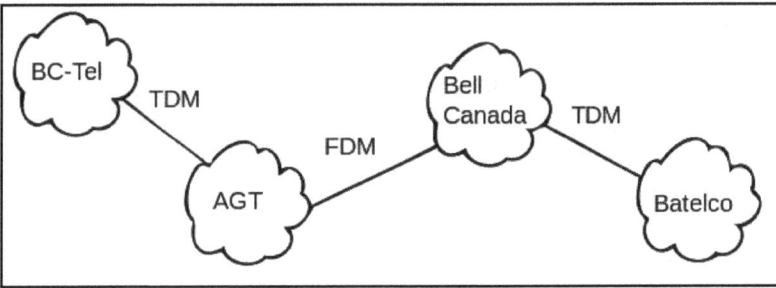

Telecommunication multiplexing.

Multiple Access Method

A multiplexing technique may be further extended into a multiple access method or channel access method, for example, TDM into time-division multiple access (TDMA) and statistical multiplexing into carrier-sense multiple access (CSMA). A multiple access method makes it possible for several transmitters connected to the same physical medium to share its capacity.

Multiplexing is provided by the Physical Layer of the OSI model, while multiple access also involves a media access control protocol, which is part of the Data Link Layer.

The Transport layer in the OSI model, as well as TCP/IP model, provides statistical multiplexing of several application layer data flows to/from the same computer.

Code-division multiplexing (CDM) is a technique in which each channel transmits its bits as a coded channel-specific sequence of pulses. This coded transmission typically is accomplished by transmitting a unique time-dependent series of short pulses, which are placed within chip times within the larger bit time. All channels, each with a different code, can be transmitted on the same fiber and asynchronously demultiplexed. Other widely used multiple access techniques are time-division multiple access (TDMA) and frequency-division multiple access (FDMA). Code-division multiplex techniques are used as an access technology, namely code-division multiple access (CDMA), in Universal Mobile Telecommunications System (UMTS) standard for the third-generation (3G) mobile communication identified by the ITU.

Application Areas

Telegraphy

The earliest communication technology using electrical wires, and therefore sharing an interest in the economies afforded by multiplexing, was the electric telegraph. Early experiments allowed two separate messages to travel in opposite directions simultaneously, first using an electric battery at both ends, then at only one end.

- Émile Baudot developed a time-multiplexing system of multiple Hughes machines in the 1870s.

- In 1874, the quadruplex telegraph developed by Thomas Edison transmitted two messages in each direction simultaneously, for a total of four messages transiting the same wire at the same time.

- Several workers were investigating acoustic telegraphy, a frequency-division multiplexing technique, which led to the invention of the telephone.

Telephony

In telephony, a customer's telephone line now typically ends at the remote concentrator box, where it is multiplexed along with other telephone lines for that neighborhood or other similar area. The multiplexed signal is then carried to the central switching office on significantly fewer wires and for much further distances than a customer's line can practically go. This is likewise also true for digital subscriber lines (DSL).

Fiber in the loop (FITL) is a common method of multiplexing, which uses optical fiber as the backbone. It not only connects POTS phone lines with the rest of the PSTN, but also replaces DSL by connecting directly to Ethernet wired into the home. Asynchronous Transfer Mode is often the communications protocol used.

Cable TV has long carried multiplexed television channels, and late in the 20th century began offering the same services as telephone companies. IPTV also depends on multiplexing.

Video Processing

In video editing and processing systems, multiplexing refers to the process of interleaving audio and video into one coherent data stream.

In digital video, such a transport stream is normally a feature of a container format which may include metadata and other information, such as subtitles. The audio and video streams may have variable bit rate. Software that produces such a transport stream and container is commonly called a statistical multiplexer or muxer. A demuxer is software that extracts or otherwise makes available for separate processing the components of such a stream or container.

Digital Broadcasting

In digital television systems, several variable bit-rate data streams are multiplexed together to a fixed bitrate transport stream by means of statistical multiplexing. This makes it possible to transfer several video and audio channels simultaneously over the same frequency channel, together with various services. This may involve several standard definition television (SDTV) programmes (particularly on DVB-T, DVB-S2, ISDB and ATSC-C), or one HDTV, possibly with a single SDTV companion channel over one 6 to 8 MHz-wide TV channel. The device that accomplishes this is called a statistical multiplexer. In several of these systems, the multiplexing results in an MPEG transport stream. The newer DVB standards DVB-S2 and DVB-T2 has the capacity to carry several HDTV channels in one multiplex.

In digital radio, a multiplex (also known as an ensemble) is a number of radio stations that are grouped together. A multiplex is a stream of digital information that includes audio and other data.

On communications satellites which carry broadcast television networks and radio networks, this is known as multiple channel per carrier or MCPC. Where multiplexing is not practical (such as where there are different sources using a single transponder), single channel per carrier mode is used.

Analog Broadcasting

In FM broadcasting and other analog radio media, multiplexing is a term commonly given to the process of adding subcarriers to the audio signal before it enters the transmitter, where modulation occurs. (In fact, the stereo multiplex signal can be generated using time-division multiplexing, by switching between the two (left channel and right channel) input signals at an ultrasonic rate (the subcarrier), and then filtering out the higher harmonics.) Multiplexing in this sense is sometimes known as MPX, which in turn is also an old term for stereophonic FM, seen on stereo systems since the 1960s.

Time-division Multiplexing

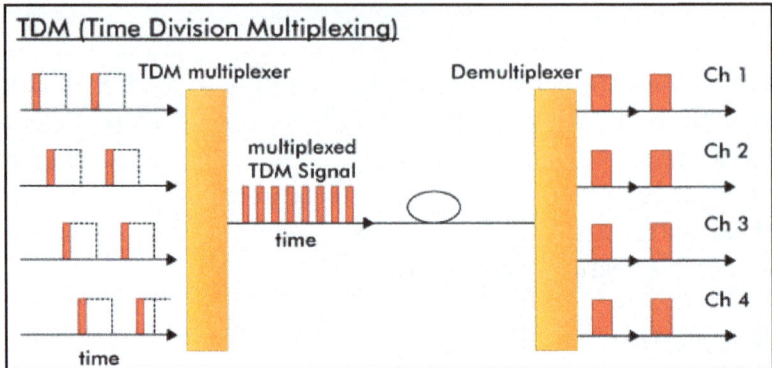

Time-division multiplexing (TDM) is a method of putting multiple data streams in a single signal by separating the signal into many segments, each having a very short duration. Each individual data stream is reassembled at the receiving end based on the timing.

The circuit that combines signals at the source (transmitting) end of a communications link is known as a multiplexer. It accepts the input from each individual end user, breaks each signal into segments, and assigns the segments to the composite signal in a rotating, repeating sequence. The composite signal thus contains data from multiple senders. At the other end of the long-distance cable, the individual signals are separated out by means of a circuit called a demultiplexer, and routed to the proper end users. A two-way communications circuit requires a multiplexer/demultiplexer at each end of the long-distance, high-bandwidth cable.

If many signals must be sent along a single long-distance line, careful engineering is required to ensure that the system will perform properly. An asset of TDM is its flexibility. The figure allows for variation in the number of signals being sent along the line, and constantly adjusts the time intervals to make optimum use of the available bandwidth. The Internet is a classic example of a communications network in which the volume of traffic can change drastically from hour to hour. In some systems, a different scheme, known as frequency-division multiplexing (FDM), is preferred.

Types of TDM

1. Synchronous TDM.

2. Asynchronous TDM.

Synchronous TDM (STDM)

1. In synchronous TDM, each device is given same time slot to transmit the data over the link, irrespective of the fact that the device has any data to transmit or

not. Hence the name Synchronous TDM. Synchronous TDM requires that the total speed of various input lines should not exceed the capacity of path.

2. Each device places its data onto the link when its time slot arrives i.e. each device is given the possession of line turn by turn.

3. If any device does not have data to send then its time slot remains empty.

4. The various time slots are organized into frames and each frame consists of one or more time slots dedicated to each sending device.

5. If there are n sending devices, there will be n slots in frame i.e. one slot for each device.

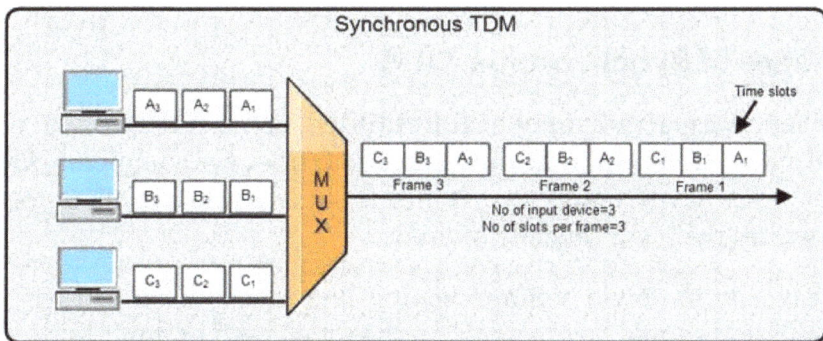

6. As show in figure, there are 3 input devices, so there are 3 slots in each frame.

Multiplexing Process in STDM

1. In STDM every device is given the opportunity to transmit a specific amount of data onto the link.

2. Each device gets its turn in fixed order and for fixed amount of time. This process is known as interleaving.

3. We can say that the operation of STDM is similar to that of a fast interleaved switch. The switch opens in front of a device; the device gets a chance to place the data onto the link.

4. Such an interleaving may be done on the basis of a hit, a byte or by any other data unit.

5. In STDM, the interleaved units are of same size i.e. if one device sends a byte, other will also send a byte and so on.

6. As shown in the fig. interleaving is done by a character (one byte). Each frame consists of four slots as there are four input devices. The slots of some devices go empty if they do not have any data to send.

7. At the receiver, demultiplexer decomposes each frame by extracting each character in turn. As a character is removed from frame, it is passed to the appropriate receiving device.

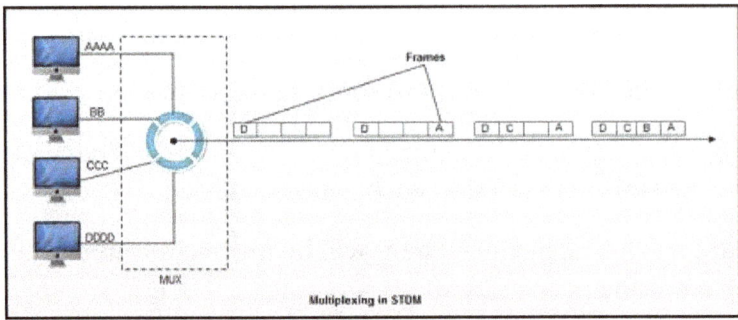

Multiplexing in STDM

Disadvantages of Synchronous TDM

1   The channel capacity cannot be fully utilized. Some of the slots go empty in certain frames. As shown in fig only first two frames are completely filled. The last three frames have 6 empty slot. It means out of 20 slots in all, 6 slots are empty. This wastes the l/4th capacity of links.

2. The capacity of single communication line that is used to carry the various transmission should be greater than the total speed of input lines.

Asynchronous TDM

1. It is also known as statistical time division multiplexing.

2. Asynchronous TDM is called so because is this type of multiplexing, time slots are not fixed *i.e.* the slots are flexible.

3. Here, the total speed of input lines can be greater than the capacity of the path.

4. In synchronous TDM, if we have n input lines then there are n slots in one frame. But in asynchronous it is not so.

5. In asynchronous TDM, if we have n input lines then the frame contains not more than m slots, with m less than n $(m < n)$.

6. In asynchronous TDM, the number of time slots in a frame is based on a statistical analysis of number of input lines.

7. In this system slots are not predefined, the slots are allocated to any of the device that has data to send.

8. The multiplexer scans the various input lines, accepts the data from the lines that have data to send, fills the frame and then sends the frame across the link.

9. If there are not enough data to fill all the slots in a frame, then the frames are transmitted partially filled.

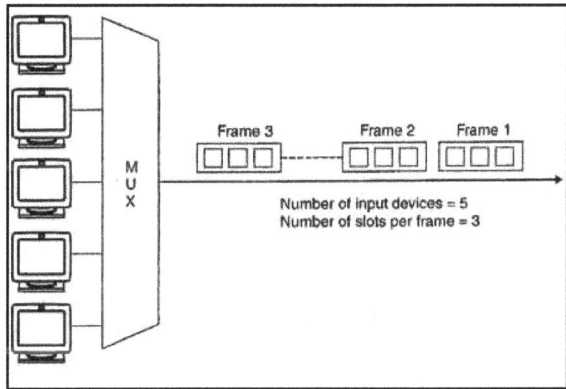

Asynchronous TDM.

10. Asynchronous Time Division Multiplexing is depicted in fig. Here we have five input lines and three slots per frame.

11. Only three out of five input lines place data onto the link *i.e.* number of input lines and number of slots per frame are same.

12. The four out of five input lines are active. Here number of input line is one more than the number of slots per frame.

13. All five input lines are active.

In all these cases, multiplexer scans the various lines in order and fills the frames and transmits them across the channel.

The distribution of various slots in the frames is not symmetrical. In case 2, device 1 occupies first slot in first frame, second slot in second frame and third slot in third frame.

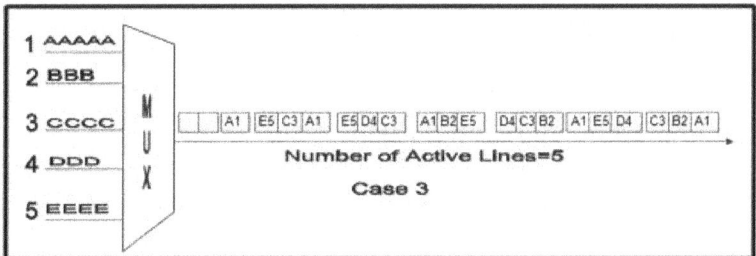

Advantages of TDM

1. Full available channel bandwidth can be utilized for each channel.

2 Intermodulation distortion is absent.

3. TDM circuitry is not very complex.

4. The problem of crosstalk is not severe.

Disadvantages of TDM

1. Synchronization is essential for proper operation.

2. Due to slow narrowband fading, all the TDM channels may get wiped out.

Application Examples

- The plesiochronous digital hierarchy (PDH) system, also known as the PCM system, for digital transmission of several telephone calls over the same four-wire copper cable (T-carrier or E-carrier) or fiber cable in the circuit switched digital telephone network.

- The synchronous digital hierarchy (SDH)/synchronous optical networking (SONET) network transmission standards that have replaced PDH.

- The Basic Rate Interface and Primary Rate Interface for the Integrated Services Digital Network (ISDN).

- The RIFF (WAV) audio standard interleaves left and right stereo signals on a per-sample basis.

TDM can be further extended into the time-division multiple access (TDMA) scheme, where several stations connected to the same physical medium, for example sharing the same frequency channel, can communicate. Application examples include:

- The GSM telephone system.

- The Tactical Data Links Link 16 and Link 22.

Multiplexed Digital Transmission

In circuit-switched networks, such as the public switched telephone network (PSTN), it is desirable to transmit multiple subscriber calls over the same transmission medium to effectively utilize the bandwidth of the medium. TDM allows transmitting and receiving telephone switches to create channels (*tributaries*) within a transmission stream. A standard DS0 voice signal has a data bit rate of 64 kbit/s. A TDM circuit runs at a much higher signal bandwidth, permitting the bandwidth to be divided into time frames (time slots) for each voice signal which is multiplexed onto the line by the transmitter. If the TDM frame consists of n voice frames, the line bandwidth is $n*64$ kbit/s.

Each voice time slot in the TDM frame is called a channel. In European systems, standard TDM frames contain 30 digital voice channels (E1), and in American systems (T1), they contain 24 channels. Both standards also contain extra bits (or bit time slots) for signaling and synchronization bits.

Multiplexing more than 24 or 30 digital voice channels is called *higher order multiplexing*. Higher order multiplexing is accomplished by multiplexing the standard TDM frames. For example, a European 120 channel TDM frame is formed by multiplexing four standard 30 channel TDM frames. At each higher order multiplex, four TDM frames from the immediate lower order are combined, creating multiplexes with a bandwidth of $n*64$ kbit/s, where $n = 120, 480, 1920$, etc.

Telecommunications Systems

There are three types of synchronous TDM: T1, SONET/SDH, and ISDN.

Plesiochronous digital hierarchy (PDH) was developed as a standard for multiplexing higher order frames. PDH created larger numbers of channels by multiplexing the standard Europeans 30 channel TDM frames. This solution worked for a while; however PDH suffered from several inherent drawbacks which ultimately resulted in the development of the Synchronous Digital Hierarchy (SDH). The requirements which drove the development of SDH were these:

- Be synchronous: All clocks in the system must align with a reference clock.

- Be service-oriented: SDH must route traffic from End Exchange to End Exchange without worrying about exchanges in between, where the bandwidth can be reserved at a fixed level for a fixed period of time.

- Allow frames of any size to be removed or inserted into an SDH frame of any size.

- Easily manageable with the capability of transferring management data across links.

- Provide high levels of recovery from faults.

- Provide high data rates by multiplexing any size frame, limited only by technology.

- Give reduced bit rate errors.

SDH has become the primary transmission protocol in most PSTN networks. It was developed to allow streams 1.544 Mbit/s and above to be multiplexed, in order to create larger SDH frames known as Synchronous Transport Modules (STM). The STM-1 frame consists of smaller streams that are multiplexed to create a 155.52 Mbit/s frame. SDH can also multiplex packet based frames e.g. Ethernet, PPP and ATM.

While SDH is considered to be a transmission protocol (Layer 1 in the OSI Reference Model), it also performs some switching functions, as stated in the third bullet point requirement listed above. The most common SDH Networking functions are these:

- *SDH Crossconnect*: The SDH Crossconnect is the SDH version of a Time Space

Time crosspoint switch. It connects any channel on any of its inputs to any channel on any of its outputs. The SDH Crossconnect is used in Transit Exchanges, where all inputs and outputs are connected to other exchanges.

- *SDH Add-Drop Multiplexer*: The SDH Add-Drop Multiplexer (ADM) can add or remove any multiplexed frame down to 1.544Mb. Below this level, standard TDM can be performed. SDH ADMs can also perform the task of an SDH Crossconnect and are used in End Exchanges where the channels from subscribers are connected to the core PSTN network.

SDH network functions are connected using high-speed optic fibre. Optic fibre uses light pulses to transmit data and is therefore extremely fast. Modern optic fibre transmission makes use of wavelength-division multiplexing (WDM) where signals transmitted across the fibre are transmitted at different wavelengths, creating additional channels for transmission. This increases the speed and capacity of the link, which in turn reduces both unit and total costs.

Statistical Time-division Multiplexing

Statistical time-division multiplexing (STDM) is an advanced version of TDM in which both the address of the terminal and the data itself are transmitted together for better routing. Using STDM allows bandwidth to be split over one line. Many college and corporate campuses use this type of TDM to distribute bandwidth.

On a 10-Mbit line entering a network, STDM can be used to provide 178 terminals with a dedicated 56k connection (178 * 56k = 9.96Mb). A more common use however is to only grant the bandwidth when that much is needed. STDM does not reserve a time slot for each terminal, rather it assigns a slot when the terminal is requiring data to be sent or received.

In its primary form, TDM is used for circuit mode communication with a fixed number of channels and constant bandwidth per channel. Bandwidth reservation distinguishes time-division multiplexing from statistical multiplexing such as statistical time-division multiplexing. In pure TDM, the time slots are recurrent in a fixed order and pre-allocated to the channels, rather than scheduled on a packet-by-packet basis.

In dynamic TDMA, a scheduling algorithm dynamically reserves a variable number of time slots in each frame to variable bit-rate data streams, based on the traffic demand of each data stream. Dynamic TDMA is used in:

- HIPERLAN/2.

- Dynamic synchronous transfer mode.

- IEEE 802.16a.

Asynchronous time-division multiplexing (ATDM), is an alternative nomenclature in which STDM designates synchronous time-division multiplexing, the older method that uses fixed time slots.

Frequency-division Multiplexing

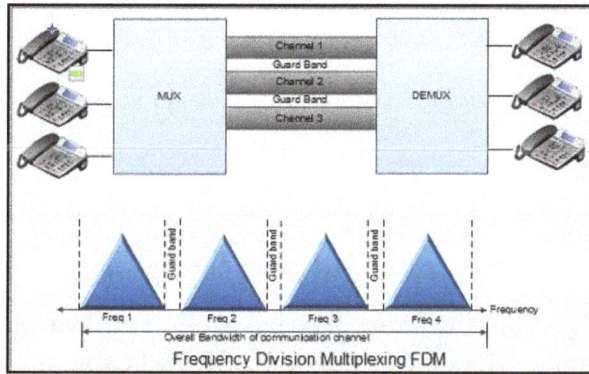

Frequency-division Multiplexing.

In frequency-division multiplexing (FDM) is a technique by which the available bandwidth of a communications channel is shared among multiple users by frequency translating, or modulating, each of the individual users onto a different carrier frequency.

Data of different end nodes are then modulated using these different carriers, so that the resultant signal of each end node occupies a different region in the frequency domain. Between each adjacent carriers, a small guard band is left unused, so as not to cause interference between closely separated carriers.

In FDM, at any instant of time, we would have electromagnetic signals corresponding to each node/sub-channel, unlike in TDM, where at any instant of time, the channel would only have electromagnetic signal belonging to one end node/sub-channel. This is shown in the diagram given below.

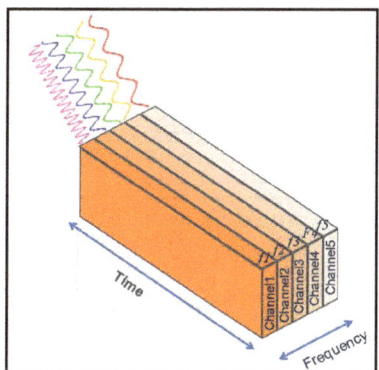

FDM with 6 different frequency carriers coexisting simultaneously in the time domain.

Traditional FM Radio and Broadcast TV are classical examples of applications using FDM, where data belonging to each radio station/TV channel is modulated over a different carrier, as shown in the diagram given below:

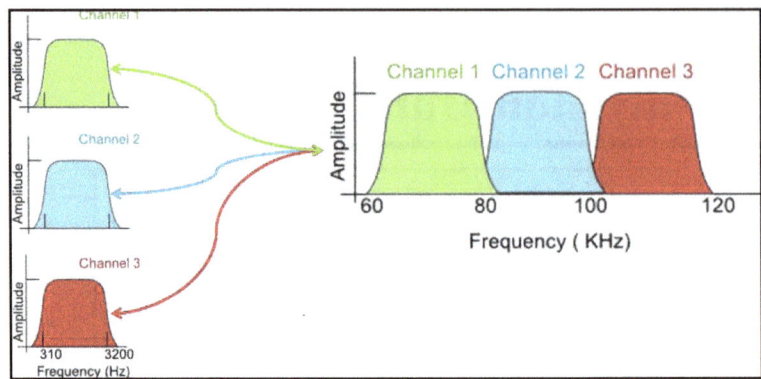

Three baseband channels modulated onto different frequency carriers to seperate them in the frequency domain.

In computer communication, the concept of basic FDM and variants of FDM are widely used both in LAN and WAN environments. DSL and cable modem links are typical examples of physical layer protocols using FDM for achieving high data rates. In DSL, which also uses the standard telephone last mile local loop line, multiple sub-carriers, each with a bandwidth of 4KHz. are used to carry users data. The baseband region from 0 to 4KHZ is left for basic POTS voice calls. Above this, some number of sub-carriers are allotted for upstream traffic and a higher number of sub-carriers are alloted for downstream traffic. Similarly, cable modem has a separate frequency band for upstream traffic and a range of sub-carriers for downstream traffic.

An example diagram showing the sub-carrier spectrum allocation for POTS, DSL upstream and downstream directions are given in the diagram below:

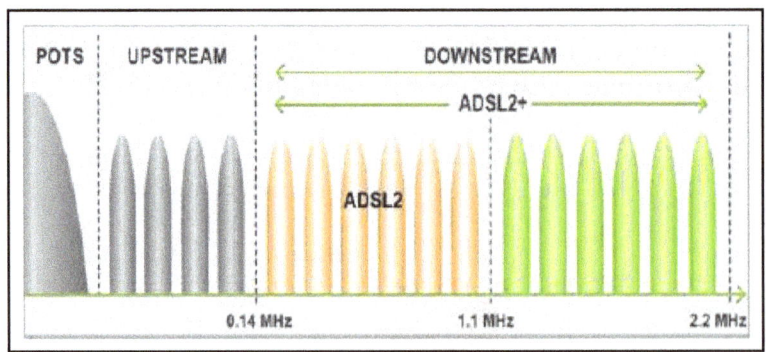

FDM being used in ADSL, with different frequency sub-carriers for POTS, ADSL upstream and ADSL downstream.

In DSL, to achieve high data rates, a line coding technique like QAM is used on top of each sub-carrier. Thus both FDM and line coding techniques are combined at the physical layer to achieve high broadband data rates.

FDM is also used in some variants of Fast Ethernet (100 Mbps) and Gigabit Ethernet (1000 Mbps) LAN protocols, where multiple carriers are used to achieve the overall data rate supported by the underlying physical layer.

Variants of FDM:

- Wavelength Division Multiplexing (WDM) and DWDM (Dense-WDM) used in optical Networks, are based on principles similar to FDM, except that their carriers are based on different wavelengths instead of different frequencies.

- Frequency Division Duplexing (FDD) is a form of FDM, where some set of frequencies/carriers are used for carrying uplink direction traffic and some other set of frequencies are used for carrying downlink traffic, thereby enabling full duplex communication using FDM.

- Spread Spectrum techniques are variants of FDM, where the data is carried or spread over a wide range of frequency spectrum. In normal FDM, a single carrier is used to carry data corresponding to an end node. But in Spread spectrum techniques, multiple carriers are used to carry data corresponding to an end node, with each carrier carrying a small piece of data. FHSS (Frequency Hopping Spread Spectrum), DSS (Direct Sequence Spread Spectrum) and OFDM (Orthogonal Frequency Division Multiplexing) are different types of spread spectrum techniques.

- In FHSS, the frequency of the carrier varies from instant to instant, whereas in DSS, data is split into smaller units and simultanesouly carried by multiple carriers, as shown in the diagram given below.

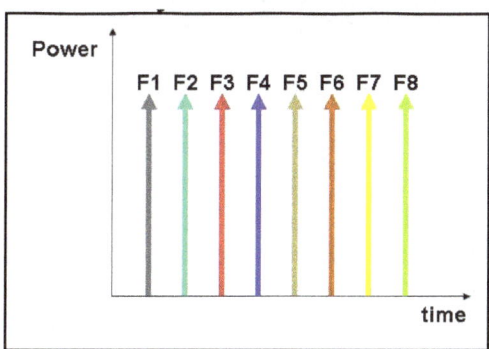

A FHSS where 8 different carriers are used, with data hopping on top of different carriers in a pre-determined order.

- CDMA (Code Division Multiple Access) is a form of DSS, where a codeword is combined with data to spread the signal over a wide range of spectrum.

- OFDM is a form of DSS that is widely used in Wireless LAN protocols (802.11 a/g), wherein a set of carriers that are orthogonal (do not interfere with each other) are used to carry the data signal, as shown in the diagram below:

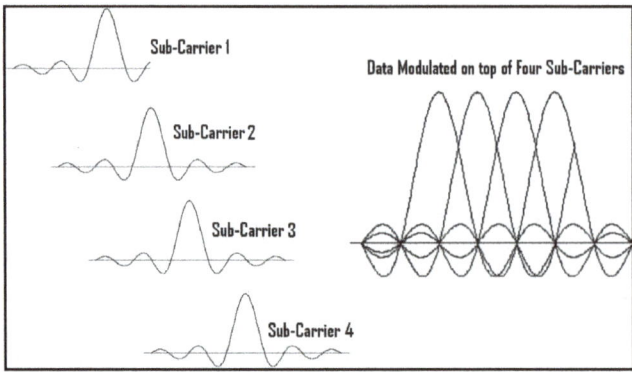

An OFDM scheme where 4 orthogonal frequency sub-carriers are
used to spread the information over a wider spectrum.

Orthogonal Frequency Division Multiplexing (OFDM) is used for both wired and wireless networks.

Working of FDM

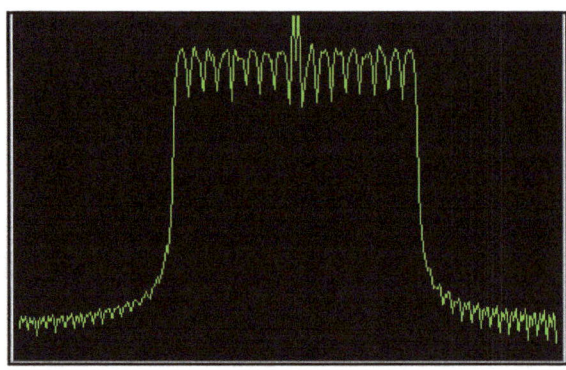

The passband of an FDM channel carrying digital data, modulated by
QPSK quadrature phase-shift keying.

The multiple separate information (modulation) signals that are sent over an FDM system, such as the video signals of the television channels that are sent over a cable TV system, are called baseband signals. At the source end, for each frequency channel, an electronic oscillator generates a *carrier* signal, a steady oscillating waveform at a single frequency that serves to "carry" information. The carrier is much higher in frequency than the baseband signal. The carrier signal and the baseband signal are combined in a modulator circuit. The modulator alters some aspect of the carrier signal, such as its amplitude, frequency, or phase, with the baseband signal, "piggybacking" the data onto the carrier.

The result of modulating (mixing) the carrier with the baseband signal is to generate sub-frequencies near the carrier frequency, at the sum $(f_C + f_B)$ and difference $(f_C - f_B)$ of the frequencies. The information from the modulated signal is carried in sidebands on each side of the carrier frequency. Therefore, all the information carried by the channel is in a narrow band of frequencies clustered around the carrier frequency, this is called the passband of the channel.

Similarly, additional baseband signals are used to modulate carriers at other frequencies, creating other channels of information. The carriers are spaced far enough apart in frequency that the band of frequencies occupied by each channel, the passbands of the separate channels do not overlap. All the channels are sent through the transmission medium, such as a coaxial cable, optical fiber, or through the air using a radio transmitter. As long as the channel frequencies are spaced far enough apart that none of the passbands overlap, the separate channels will not interfere with each other. Thus the available bandwidth is divided into "slots" or channels, each of which can carry a separate modulated signal.

For example, the coaxial cable used by cable television systems has a bandwidth of about 1000 MHz, but the passband of each television channel is only 6 MHz wide, so there is room for many channels on the cable (in modern digital cable systems each channel in turn is subdivided into subchannels and can carry up to 10 digital television channels).

At the destination end of the cable or fiber, or the radio receiver, for each channel a local oscillator produces a signal at the carrier frequency of that channel, that is mixed with the incoming modulated signal. The frequencies subtract, producing the baseband signal for that channel again. This is called demodulation. The resulting baseband signal is filtered out of the other frequencies and output to the user.

Telephone

For long distance telephone connections, 20th century telephone companies used L-carrier and similar coaxial cable systems carrying thousands of voice circuits multiplexed in multiple stages by channel banks.

For shorter distances, cheaper balanced pair cables were used for various systems including Bell System K- and N-Carrier. Those cables didn't allow such large bandwidths, so only 12 voice channels (double sideband) and later 24 (single sideband) were multiplexed into four wires, one pair for each direction with repeaters every several miles, approximately 10 km. By the end of the 20th Century, FDM voice circuits had become rare. Modern telephone systems employ digital transmission, in which time-division multiplexing (TDM) is used instead of FDM.

Since the late 20th century digital subscriber lines (DSL) have used a Discrete multitone (DMT) system to divide their spectrum into frequency channels.

The concept corresponding to frequency-division multiplexing in the optical domain is known as wavelength-division multiplexing.

Group and Supergroup

A once commonplace FDM system, used for example in L-carrier, uses crystal filters which operate at the 8 MHz range to form a Channel Group of 12 channels, 48 kHz bandwidth in the range 8140 to 8188 kHz by selecting carriers in the range 8140 to 8184 kHz selecting upper sideband this group can then be translated to the standard range 60 to 108 kHz by a carrier of 8248 kHz. Such systems are used in DTL (Direct To Line) and DFSG (Directly formed super group).

132 voice channels (2SG + 1G) can be formed using DTL plane the modulation and frequency plan are given in FIG1 and FIG2 use of DTL technique allows the formation of a maximum of 132 voice channels that can be placed direct to line. DTL eliminates group and super group equipment.

DFSG can take similar steps where a direct formation of a number of super groups can be obtained in the 8 kHz the DFSG also eliminates group equipment and can offer:

- Reduction in cost 7% to 13%.

- Less equipment to install and maintain.

- Increased reliability due to less equipment.

Both DTL and DFSG can fit the requirement of low density system (using DTL) and higher density system (using DFSG). The DFSG terminal is similar to DTL terminal except instead of two super groups many super groups are combined. A Mastergroup of 600 channels (10 super-groups) is an example based on DFSG.

Examples

FDM can also be used to combine signals before final modulation onto a carrier wave. In this case the carrier signals are referred to as subcarriers: an example is stereo FM transmission, where a 38 kHz subcarrier is used to separate the left-right difference signal from the central left-right sum channel, prior to the frequency modulation of the composite signal. An analog NTSC television channel is divided into subcarrier frequencies for video, color, and audio. DSL uses different frequencies for voice and for upstream and downstream data transmission on the same conductors, which is also an example of frequency duplex.

Where frequency-division multiplexing is used as to allow multiple users to share a physical communications channel, it is called frequency-division multiple access (FDMA).

FDMA is the traditional way of separating radio signals from different transmitters.

In the 1860s and 70s, several inventors attempted FDM under the names of acoustic telegraphy and harmonic telegraphy. Practical FDM was only achieved in the electronic age. Meanwhile, their efforts led to an elementary understanding of electroacoustic technology, resulting in the invention of the telephone.

Space-division Multiplexing

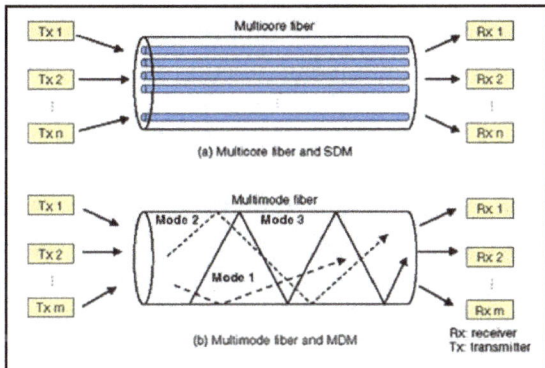

Space-division multiplexing is a method by which metallic, radio, or optical transmission media are physically separated by insulation, waveguides, or space in order to maintain channel separations. Within each physically distinct channel, multiple channels can be derived through frequency, time, or wavelength division multiplexing. Some Passive Optical Network (PON) implementations employ space division multiplexing, with the downstream transmissions occurring over one fiber of a duplex fiber optic cable and upstream transmission occurring over the other fiber.

Space Division Multiplexing in Optical Fibres

The notion of increasing fibre capacity with Space Division Multiplexing (SDM) is almost as old as fibre communications itself, with the fabrication of fibres containing multiple cores, the first and most obvious approach to SDM, reported as far back as 19791. Yet only recently has serious attention been given to building a complete networking platform as needed to make use of this multicore fibre (MCF) approach. The alternative approach of using modes within a multimode fibre (MMF) as a means to define separate spatially distinct channels also dates back to that era2.

The current frenzied progress in SDM is occurring now because of a convergence of enabling technological capabilities and a rapidly emerging need. On the one hand, SDM draws on the accumulated progress of fibre research. This includes subtle improvements in traditional fibres3, and the fantastically precise fabrication methods developed to produce hollow-core and other complex microstructure fibres. Sophisticated mode control8 and analysis9 methods along with tapered devices can be borrowed from high-power fibre laser research, which itself has needed to develop means to better exploit the spatial domain in the drive to achieve ever higher power levels. Photonic lantern and endoscope devices are available from their development for imaging.

Today's SDM research is also occurring as coherent detection and digital compensation are capable of overcoming complex impairments (such as polarization mode dispersion (PMD)) and are accepted as a standard part of high-performance systems. This is crucial: since SDM packs spatial channels tightly into each fibre, crosstalk between channels is an obvious potential disadvantage and needs to be addressed. The addition of significant crosstalk to a transmission line would have been particularly unattractive a few years ago, before coherent-detection systems offered hope of subtracting out crosstalk electronically at the receiver.

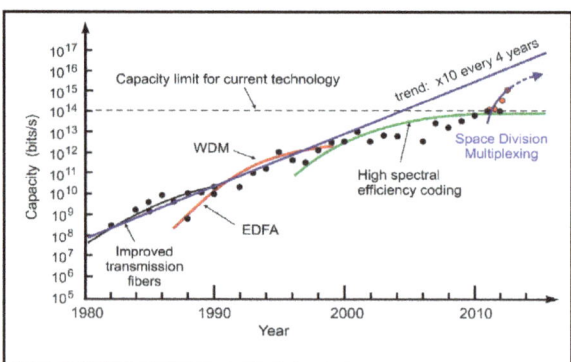

Figure: The evolution of transmission capacity in optical fibres as evidenced by state of the art laboratory transmission demonstrations over the years.

The data points shown represent the highest capacity transmission numbers (all transmission distances considered) as reported in the Postdeadline Session of the Optical Fiber Communications conference held each year in the USA. The transmission capacity of a single fibre strand is seen to have increased by approximately a factor of 10 every 4 years. Key previous technological breakthroughs include the development of low-loss single-mode fibres, the Erbium Doped Fibre Amplifier (EDFA), Wavelength Division Multiplexing (WDM) and more recently high-spectral efficiency coding via DSP-enabled coherent transmission. The data points for Space Division Multiplexing also include results from the Postdeadline Session at the annual European Conference on Optical Communications (ECOC). As can be seen SDM appears poised to provide the next step change in transmission capacity.

These enabling technologies have made SDM a viable strategy just as a severe need for innovation emerges. Over the past forty years, a series of technological breakthroughs have allowed the capacity-per-fibre to increase around 10x every four years, as illustrated in figure above. Transmission technology has therefore thus far been able to keep up with the relentless, exponential growth of capacity demand. The cost of transmitting exponentially more data was also manageable, in large part because more data was transmitted over the same fibre by upgrading equipment at the fibre ends. But in the coming decade or so, an increasing number of fibres in real networks will reach their capacity limit14. Keeping up with demand will therefore mean lighting new fibres and installing new cables - potentially also at an exponentially increasing rate. Further, this fibre capacity limit is not specific to a particular modulation format or transponder standard - it is fundamental and can be derived from a straightforward extension of the fundamental Shannon capacity limit to a nonlinear fibre channel under quite broad assumptions15. It says that standard single mode fibre (SMF) can carry no more than around 100Tbit/s of data, corresponding to filling the C and L amplification bands of the erbium doped fibre amplifier (EDFA) at a spectral efficiency of ~ 10 bits/ s/ Hz .

The upcoming potential "capacity crunch", then, is an era of unfavourable cost scaling. For some carriers who have access to a limited number of dark fibres, very expensive installation of new cables will be the only alternative as the capacity of existing fibres is filled. "Fibre-rich" carriers who attempted to future-proof their fibre plant by including large numbers of premium fibres in each cable (thus putting off the need for subsequent new cables) will be forced to overbuild, i.e. deploy multiple systems over parallel fibres to keep up with demand. However, multiple systems over parallel fibres suggest that transmission costs and power consumption will scale linearly with growing capacity. The fear is that, without further innovation to lower the cost-per-bit, the capacity crunch will apply pressure to constrain growth, and we will finally reach the end of the seemingly boundless connectivity that drives our economy and enriches our experiences.

The anticipated promise of SDM is not only that it will provide the next leap in capacity-per-fibre, as shown in figure, but that this will concurrently enable large reductions in cost-per-bit and improved energy efficiency16. This is a formidable challenge. SDM is very different from wavelength division multiplexing (WDM) which inherently allows the sharing of key components: e.g., an EDFA and dispersion compensation module can easily be shared by many WDM channels with minimal added complexity. The benefits of SDM are more speculative, and assume that many system components can be eventually integrated and engineered to support this potentially disruptive new platform.

Given this emerging need, major research effort has been mobilized around the world to explore and establish the viability of SDM. Exciting recent results show that a wide array of new tools are now being focused on probing the potential benefits of SDM, and chipping away at the many engineering problems obscuring these benefits.

Technical Approaches to SDM

The term SDM is nowadays taken to refer to multiplexing techniques that establish multiple spatially distinguishable data pathways through the same fibre, although in earlier days the same terminology was previously applied to describe the case of multiple parallel fibre systems: the benchmark that needs to be beaten on a cost-per-bit perspective if any of the SDM approaches currently under investigation are ever to be commercially deployed. The primary technical challenge given the more intimate proximity of the pathways is management of cross-talk.

In the case of multicore fibre (MCF) in which the distinguishable pathways are defined by an array of physically-distinct single-mode cores (Figure b) the simplest way to limit cross-talk is to keep the fibre cores well-spaced. Small variations in core properties, either deliberately imposed across the fibre cross-section, or due to fabrication/cabling19, can also reduce cross-coupling along the fibre length. As will be discussed later, to date, the highest capacities and longest transmission distances demonstrated in SDM system experiments have all utilized such "uncoupled" MCFs. A study of the tolerance of various advanced modulation format signals to in-band accumulated cross- talk (to include contributions from signal multiplexing and demultiplexing, amplification, splicing and distributed-coupling along the fibre length) showed that < -25dB cross-talk levels are typically required to avoid significant transmission penalties20.

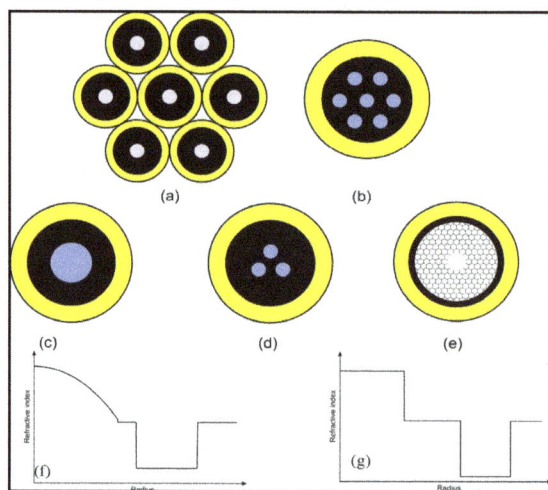

The different approaches to SDM.

(a) Fibre-bundles composed of physically-independent, single-mode fibres of reduced cladding dimension could provide for increased core packing densities relative to current fibre cables, however "in-fibre" SDM will be needed to achieve the higher core densities and levels of integration ultimately desired.

(b) MCF comprising multiple independent cores sufficiently spaced to limit cross-talk. Fibres with up to 19 cores have so far been demonstrated for long haul transmission– higher core counts are possible for short haul applications (e.g. datacomms) which can tolerate higher levels of cross-talk per unit length.

(c) FMF with a core dimension/numerical aperture set to guide a restricted number of modes – so far typically 6-12 distinct modes (including all degeneracies and polarisations). To date work has focussed primarily on using the first few LP-fibre modes; however, work is now beginning on using other modal basis sets that exploit the true vector modes of the fibre – in particular on modes that carry orbital angular momentum and which may provide benefits in terms of reduced mode-coupling and associated DSP requirements.

(d) Coupled-core fibres support super mode that allow for higher spatial mode densities than isolated-core fibres. MIMO processing is essential to address the inherent mode-coupling.

(e) Photonic Band Gap fibres4,5 guide light in an air-core and thus have ultra-low optical nonlinearity, offer the potential for lower losses than solid core fibres (albeit at longer transmission wavelengths around 2μm rather than 1.55μm 43,44). Work is underway to understand whether such fibres can support MDM and to establish their practicality for high capacity communications.

(f) Refractive index profile of a GI core design providing low DMGD and low mode-coupling for long haul FMF transmission.

(g) Core refractive index design incorporating a trench profile to reduce cross-talk and thus allow closer core separations in MCF.

Using trench-type core refractive index profiles matched to standard SMF, to better confine the mode, it has proved possible to reduce core-to-core coupling to impressively low levels (<-90dB/km) for a spacing of around 40μm, enabling transmission over multi-1000km length scales. However, fibre reliability issues, in particular susceptibility to fracture, mean that MCF diameters beyond 200μm are not considered practical, placing a fairly firm bound on the number of cores that can be incorporated in MCFs for long-haul transmission. Most fibres to date have used a hexagonal arrangement of 7-cores for which the central core, with six nearest-neighbours as opposed to three for cores in the outer ring, suffers the highest level of cross-talk. More recent work has used 12-cores arranged in a ring geometry such that each core has just two nearest-neighbours and experiences nominally the same level of cross-talk (-57 dB/km in this case). A 19- core fibre of 200μm outer diameter has also been reported; however the cross-talk was already substantially higher and limited the useful transmission distance to ~10km.

The situation is quite different for mode division multiplexed (MDM) transmission in MMF where the distinguishable pathways have significant spatial overlap and, as a consequence, signals are prone to couple randomly between the modes during propagation. In general the modes will exhibit differential mode group delays (DMGD) and also differential modal loss or gain. The energy of a given data symbol launched into a particular mode spreads out into adjacent symbol time slots as a result of mode-coupling, rapidly compromising successful reception of the information it carries. Crosstalk occurs when light is coupled from one mode to another and remains there upon detection. Inter-symbol interference occurs when the crosstalk is coupled back to the original mode after propagation in a mode with different group velocity. As in wireless systems, equalization utilizing multiple-input multiple-output (MIMO) techniques[24] is required at the receivers to mitigate these linear impairments.

MIMO signal processing is already widely used in current coherent optical transmission systems with polarization division multiplexing (PDM) over standard single-mode fibres. A 2x2 realization with four finite impulse response (FIR) filters recovers the signals on the two polarizations and compensates for PMD. For an MDM system with M modes, the respective algorithms would need to be scaled to 2Mx2M MIMO, requiring 4M adaptive FIR filters. (By way of comparison, the same capacity carried on M uncoupled SDM waveguides would require 4M adaptive FIR filters.) Thus, if we assume an equal number of taps per adaptive FIR filter and equal complexity of the adaptation algorithm, comparing a 2M×2M MIMO system on M coupled waveguides to M uncoupled SDM waveguides using PDM results in a complexity scaling as 4M/(4M) = M. To compensate DMGD and mode cross-talk completely, the equalization filter length should be larger than the impulse response spread. The computational complexity of FIR filters implemented as time-domain equalizers (TDE) increases linearly with the total DMGD of the link, which can make TDE unfeasible for long-haul MDM transmission. Common equalizer algorithms were studied and orthogonal frequency division multiplexing (OFDM) was found to achieve the lowest complexity[27]. However, for OFDM the DMGD to be compensated (and thus the reach) is limited by the length of the cyclic prefix. In other work aimed at lowering the DSP complexity, single-carrier adaptive frequency-domain equalization (SC-FDE) for MDM transmission has recently been proposed, where the complexity of SC-FDE scales logarithmically with the total DMGD.

Conventional MMFs with core/cladding diameters of 50/125 and support more than 100 modes and have large DMGDs, and thus are not suitable for long-haul transmission because the DSP complexity would be too high. Recent advancements have led to fibres supporting a small number of modes, the so-called "few-mode fibres" (FMFs), with low DMGD. The most significant research demonstrations have so far concentrated on the simplest FMF, which supports three modes, the LP01 and degenerate LP11 modes, for a total of 6 polarization and spatial modes. The DMGD in step-index core designs (as used in the first demonstrations of MDM in 3MF) is a few ns/km, meaning that the number of taps required for MIMO processing was impractical for transmission distances much greater than 10km. Consequently work has been undertaken to develop core designs offering substantially reduced values of DMGD. Using a graded-index (GI-) core design, DMGD values as low as 50 ps/km have been achieved for 3MF. Moreover, it has been shown that DMGD cancellation is possible by combining fibres fabricated to have opposite signs of DMGD. In this way transmission lines with net values of DMGD as low as ~ 5 ps/km (and with low levels of inherent mode-coupling between mode-groups) have been realised enabling transmission over >1000km length scales when incorporated with an appropriate amplification approach[33]. Whilst these results are in themselves technically impressive, the question arises as to how scalable the basic approach will ultimately prove. To this end, experiments have been undertaken on both 6-mode[34] and 5-mode FMF[35] with encouraging initial results obtained. However, just as with the MCF approach it is clear that scaling MDM much beyond this is likely to prove very challenging, not least in terms of developing scalable, accurate, low-loss mode launch schemes and ensuring that the required DSP remains tractable. Note that in the future, MCF-based systems may also be designed to use MIMO equalization to deal with increased crosstalk arising from higher core densities and/or highly integrated transmitters/receivers, optical amplifiers, and switching elements.

Whilst zero crosstalk would be ideal, there is a developing school of thought that contends that mode-coupling is inevitable, that full 2Mx2M MIMO is thus necessary, and that strong coupling should be actively exploited. If mode-coupling is weak then a data symbol carried by multiple modes with different group indices will spread in time linearly with fibre length. In contrast, if the coupling is strong, then the temporal spread follows a random-walk process, and will scale with the square-root of fibre length. Strong coupling can therefore potentially reduce the number of MIMO taps required and consequently the DSP complexity. Indeed this is analogous to spinning of current single-mode fibre during fabrication to reduce PMD. Similarly, the impact of differential modal gain and loss can in principle be mitigated by strong mode-coupling over a suitable length scale relative to the amplifier spacing. In the MCF case, by bringing the cores closer together to ensure strong mode-coupling, it is possible to establish super mode defined by the array of cores, which can then be used to provide spatial information channels for MDM to which MIMO can be applied. This enables higher spatial channel densities for MCFs than can be obtained using isolated cores designs.

SDM Technology and Integration

While development of innovative fibres for SDM goes on, researchers have turned to component and connectivity challenges that are essential to building systems around SDM fibres. Recent component demonstrations give us glimpses of what is possible, but are based on varied and often conflicting assumptions about the larger system. Two visions of a total system are of particular interest: the "grand vision" of an ultra-high capacity, fully-SDM system, and the "upgrade-path" vision, where SDM components and links are gradually added to existing non-SDM infrastructure.

(a) Elegantly scalable passive multiplexers,

(b) Integrated transmitter and receiver arrays providing low-loss coupling to many modes of an SDM fibre,

(c) Reconfigurable routing elements that can direct SDM traffic without the need for electronic MIMO in between transmitter and receiver. The strategy for switching wavelengths and modes, as well as any additional required functionality, will determine the complexity of the ROADM architecture.

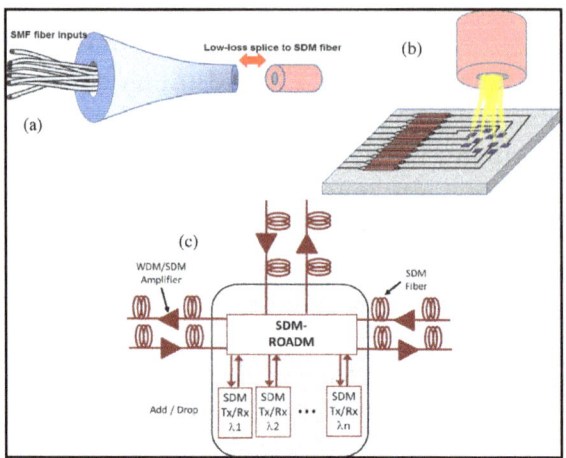

Many components are needed to fully utilize the benefit of high-density SDM, including.

When building components for the fully-SDM vision, we look into an eventual future where high performance MIMO is available, and where the density of spatial multiplexing must be increased well beyond the low-crosstalk limits. Scalability takes primary importance, whereas low-crosstalk designs confer no benefit. These broad guidelines suggest what component types will be preferred: the constituent modulators or detectors in a transmitter or receiver should be coupled to an SDM fibre in a very scalable way without regard for any one-to-one mapping of signals to modes (since the modes will quickly scramble in any case). Such scalability is natural, for example, in photonic-lantern multiplexers illustrated in figure above and in spot-type transmitters illustrated in figure above. A real system will use passive multiplexers wherever possible, to avoid cost and power dissipation. Flexible devices for actively multiplexing a desired spatial channel, such as those based on spatial light modulators, may be important, but only in the small number of subsystems where reconfigurable optical add/drop functions for spatial channels are needed.

Amplifier design is somewhat less clear in this fully-SDM regime. The potential cost and power savings of a cladding-pumped architecture are quite attractive. On the other hand, demonstrations of few-mode core-pumped amplification with low differential gain are interesting and may perhaps be scaled to larger number of modes. In either case, further work will be needed to develop gain fibres that can achieve high efficiency and low differential gain for many modes.

Today's flexible photonic mesh networks are based on reconfigurable add-drop multiplexers (ROADM), which provide carriers the ability to remotely establish lightpaths and efficiently switch those lightpaths on demand. Similar routing flexibility is assumed in future fully-SDM networks; however, there are many options for how this could be attempted. Figure (c) shows an example of a possible SDM network node, consisting of an SDM ROADM and associated SDM-WDM transmitters and receivers for the add and drop, along with SDM fibres and amplifiers for the three fibre directions addressed. The SDM ROADM itself could be quite complicated depending on the choices made regarding the type of super channels (e.g. frequency or spatial) and the strategy for switching, for example:

1) Wavelength by wavelength, all modes together;

2) Mode by mode, with mode interchange;

3) Any wavelength/mode to any other wavelength/mode. The SDM-ROADM architecture also will depend on its support for colorless, non-directional, contentionless, and gridless operation.

The general strategy that mode-coupling is inevitable and should be sorted out only at the receiver suggests that mode-independent routing elements will be a key technology: these route wavelengths to different destinations while keeping all spatial modes together. All modes will then be present at the receiver, allowing MIMO processing to recover the transmitted signals. A first demonstration of a few-mode-compatible OADM. Spatial superchannels and the plausibility of joint digital signal processing have also been considered. Cvijetic give a forward-looking view of SDM as a tool for ultimate flexibility in routing. As it matures, SDM may offer benefits beyond high capacity and flexible routing, for example in secure data transmission.

In a very different vision of SDM, upgrades of an existing single-mode infrastructure are incorporated incrementally. They must be reverse-compatible and offer cost advantages in the short term. The key here is not scalability, but compatibility, and so low-crosstalk solutions are extremely important: they allow SDM fibres and components to be used as drop-in replacements for their non-SDM counterparts (leading to a "hybrid SDM-SMF" network), and need not wait until real-time implementations of MIMO processing have been commercialized. This vision does not contradict the assumptions of the "grand vision," so much as it looks at a different time scale. Initial steps down an SDM "upgrade path" could occur quite soon, including, for example, replacing cables in a few ducts of unusually high congestion, without changing the surrounding network. Seamless connectivity will be a key requirement in early upgrades. Basic passive multiplexers have been demonstrated using free-space and tapered approaches and both are already rapidly progressing towards more practical and compact solutions.

Upgrading a conventional SMF system with SDM amplifiers could offer considerable advantages if multicore amplifiers achieve the efficiency improvements potentially available from cladding-pumping and pump sharing. This would provide system companies an incremental upgrade with a clear motivation, but is technically challenging. Initial steps towards a cladding- pumped MCF-EDFA are interesting, but are far from achieving high efficiency.

A key motivation for SDM is its potential to facilitate integration, and this will be a priority both in incremental upgrades and later progress towards fully SDM systems. Synergistic development of integrated transmitter and receiver arrays along with compatible fibres and components is essential, and was first seen in short-reach data communications systems. Silicon integrated devices matched to fibre with multiple single-mode cores have been demonstrated with seven cores and eight cores. Devices for producing the complex orthogonal fields matching specific modes of a MMF have also been demonstrated. Many other functions are needed in a real system. For example, a Raman pump sharing circuit suggests how component counts in transmission equipment can be greatly reduced by integration. The net benefit of integration will be clearer as engineering of interconnects and multiplexers brings total loss in line with single-mode devices.

Progress in Systems Demonstrations

Initial transmission experiments over MCF were aimed at short-reach applications. Following a proposal to make high density cables for FTTH applications, simultaneous 850nm, 1Gb/s transmission over two cores of a four-core MCF. More recently, 1310nm and 1490nm signals were transmitted over an 11.3-km seven-core MCF for passive optical network applications, where a fibre-based tapered multi-core connector (TMC) was first utilized to couple signals into and out of the MCF. For optical data link applications, seven- core multimode MCFs fabricated from graded-index core-rods were also demonstrated in 2010, with transmission 7 x 10-Gb/s over 100m using TMCs and discrete 850nm VCSELs. Lee reported a hexagonal array of VCSELs for coupling directly to the outer six cores of the seven-core multimode MCF, and subsequently developed a vertically illuminated photodiode array matched to the MCF, thus demonstrating a full 100m MCF optical link at up to 120Gb/s.

Following significant efforts on the design and fabrication of single-mode MCFs, demonstrations of SDM transmission over MCF for long-haul applications have shown impressive progress in terms of capacity, reach, and spectral efficiency, as detailed in table. The first WDM transmission experiments over MCFs were simultaneously reported by two groups using seven-core MCF, with 56Tb/s capacity over 76.8km and 109Tb/s capacity over 16.8km. In both experiments, the MCF cross-talk was sufficiently low such that the signals on each core could be received independently (without MIMO processing), and optical amplification utilized conventional single-core EDFAs placed before and after the MCF. Over the past two years, a number of subsequent experiments over seven-core MCFs have been performed with spectral efficiencies of 14 bit/s/Hz or higher.

Year	Reference	Fiber Type	Cores/ modes	Distance (km)	Span length (km)	Channel Rate (Gb/s)	WDM channels in each core/ mode	Net Total Efficiency (b/s/ Hz)	Net Total Capacity (Tb/s)	Capacity Distance Product (Pb/s* km)
2012	22	MCF	12	52	52	456	222	91.40	1012.32	52.64
2012	82	MCF	7	6160	55	128	40	14.44	28.88	177.87
2012	81	MCF	7	845	76.8	603	8	42.20	33.77	28.53
2012	23	MCF	19	10.1	10.1	172	100	30.50	305	3.08
2012	96	µstr MCF	3	4200	60	80	5	3.84	0.96	4.03
2011	79	MCF	7	2688	76.8	128	10	15.02	7.51	20.19
2011	80	MCF	7	76.8	76.8	1120	1		7.84	0.60
2011	41	µstr MCF	3	1200	60	80	1		0.22	0.27
2011	40	coupl MCF	3	24	24	56	1		0.15	0.004
2011	58	MCF	7	76.8	76.8	107	160	14.00	112.00	8.60
2011	76	MCF	7	76.8	76.8	107	80	14.00	56.00	4.30
2011	77	MCF	7	16.8	16.8	172	97	11.25	109.14	1.83
2012	99	MCF with SM and Fm cores	12 SM cores, 2 FM cores	3	3	1050	385 in Sm cores, 354 in FM cores	109.00	1050.00	3.15
2012	34	FMF	6	130	65	80	8	7.68	3.07	0.40
2012	32	FMF	3	119	119	256	96	12.00	57.60	6.85
2012	94	FMF	3	209	209	80	5	4.42	1.10	0.23
2012	33	FMF	3	1200	30	80	1		0.19	0.23
2012	93	FMF	3	85	85	112	1		0.31	0.03
2012	91	FM F	3	96	96	80	1		0.22	0.02
2011	35	FMF	5	40	40	112	1		0.52	0.02
2011	92	FMF	3	50	50	112	88	6.19	27.23	1.36
2011	33	FMF	3	33	33	112	6	6.19	1.86	0.06
2011	90	FMF	3	10	10	56	1		0.15	0.002
2011	89	FMF	2	40	40	112	1		0.21	0.01
2011	88	FMF	2	4.5	4.5	107	1	5.40	0.20	0.001

Table: Summary of progress in SDM system experiments.

Upper rows show experiments utilizing multicore fibres (MCF). Center row shows a MCF-FMF result. Lower rows show transmission results over few-mode fibres (FMF). The channel rate includes polarization multiplexing; the Net Spectral Efficiency and Net Total Capacity exclude the over-head for forward-error-correction.

SM: single-mode, FM: few-mode, coupled MCF: coupled-core MCF, MCF: microstructured, coupled-core MCF. The majority of SDM transmission experiments have utilized either 7-core MCF or FMF supporting 3 spatial modes. The table shows the rapid progress in both reach and net capacity achieved in just two years. Recent experiments utilizing a 12-core MCF and a MCF with hybrid SM and FM cores have demonstrated record net capacities of more than 1 Pb/s.

Several experiments have utilized coherent optical orthogonal frequency-division multiplexed (CO-OFDM) superchannels to demonstrate ultra-high per-channel bit rates over MCF. Liu and co-workers transmitted a single 1.12Tb/s superchannel comprised of 20 OFDM subchannels over a single 76.8km span of MCF80. A later WDM experiment demonstrated transmission of eight 603Gb/s superchannels over 845 km and achieved 42.2 bit/s/Hz spectral efficiency81.

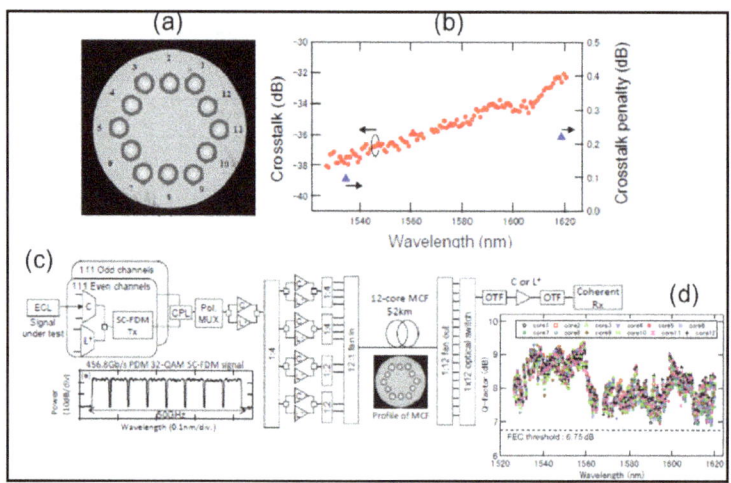

Figure: 1.01 Pbit/s MCF WDM/SDM/PDM transmission experiment.

(a) Microscope image of the cross section of the one-ring, 12-core fibre.

(b) Total crosstalk from all other cores after transmission over the 52km 12-core MCF and fan-in/fan-out devices   (filled circles), and crosstalk penalty measured for two channels (filled triangles).

(c) Schematic diagram of the transmission system   setup. ECL: external cavity laser, SC-FDM. Tx: single-carrier frequency-division multiplexed transmitter, CPL: coupler, Pol Mux: polarization multiplexer, OTF: optical tunable filter, Rx: receiver.

(d) Measured Q-factors of the 222 WDM channels in each of the   12 cores after 52-km transmission.

The longest transmission distance over MCF was recently reported by Takahashi and co- workers82, who transmitted forty 103Gb/s channels over 6160 km. That experiment was the first to utilize a multi-core EDFA for long-haul, WDM transmission over MCF, and a record capacity- distance product of 177 (Pb/s) was achieved. A record capacity of

1.01Pb/s was transmitted over a 52km MCF with 12 cores arranged in a single ring22, as shown in Figure (a). This MCF and its fan-in and fan-out devices were specifically designed for sufficiently low cross-talk to allow high-order modulation formats20, in this case 32-QAM at 400Gb/s per channel. In addition to supporting an aggregate spectral efficiency of 91.4 bit/s/Hz, the MCF had sufficiently low loss and crosstalk across the C- and extended L-bands (1526 to 1620nm), as displayed in figure above(b), thus enabling transmission of 222 channels to achieve the 1.01 Pb/s capacity (Figures above (c,d)).

The MDM concept was first proposed in 1982 when Berdague and Facq used spatial filtering techniques to launch and detect two modes at the ends of a 10m conventional graded-index MMF. It was not until 2000 when Stuart recognized the analogy between wireless and optical channels and demonstrated the application of 2x2 MIMO for reception of two MDM channels after 1km transmission. Coherent optical 2x2 MIMO was demonstrated at 800Mb/s over 100m and 2.8km of 62.5-µm MMF84. Additional proof-of-principle demonstrations utilized direct-detection and 2x2 MIMO85, and modal-diversity with direct-detection using 2x4 MIMO. These early transmission experiments all over conventional MMF did not launch and receive all modes, and thus would not have been capable of achieving the low-outage expected of optical communication.

Recent FMF development has enabled rapid progress in the capacity and reach of MDM systems demonstrations, as shown in table. In early 2011, three experiments over the simplest FMF supporting the LP01 and degenerate LP11 modes were reported at the same conference. Per- channel rates of 100Gb/s over two spatial modes were achieved over 4.5km88 and 40km89, while 56Gb/s signals in three modes were transmitted over 10km90 (the quoted bit-rates include polarization multiplexing). The three-mode experiment first demonstrated full use of all degrees of freedom afforded by the FMF (6 spatial and polarization modes), with signal recovery via full coherent 6x6 MIMO. Randel subsequently demonstrated WDM, MDM and PDM transmission of 6 wavelengths, 3 modes, and 2 polarizations, achieving 2.02Tb/s capacity (before subtracting the FEC overhead).

In the first transmission experiment to include amplification in a MMF, a few-mode inline EDFA boosted the 88 WDM signals before the mode demultiplexer and reception. To reduce mode-dependent gain, the MM-EDFA was forward pumped in the LP21 mode and reverse pumped in the LP11 mode. In later work, the transmission distance was increased to 85km. Distributed Raman amplification was employed to demonstrate single-span MDM transmission over 209km. In that experiment, the FMF supporting 6 spatial and polarization modes consisted of a first section of large effective area (>155um²) depressed-cladding FMF (which is tolerant to nonlinear effects) followed by a graded-index FMF with smaller effective area (<67um²) for efficient Raman pumping. The DMGD was compensated by using fibres spools with DMGD of opposite sign to reduce the required number of equalizer taps in the MIMO processing.

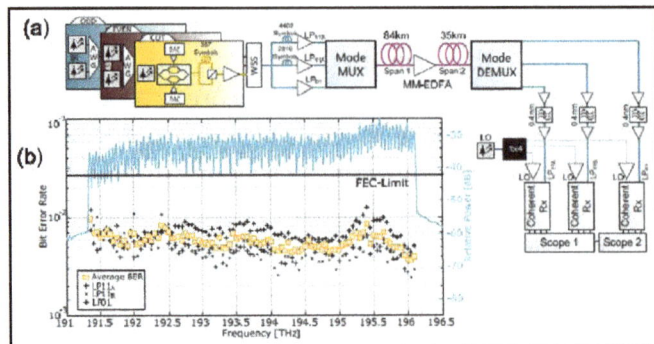

Tbit/s amplified WDM/MDM/PDM transmission experiment over a few-mode fibre[32].

(a) Schematic of the experimental setup, showing the mode multiplexer and demultiplexer and simultaneous reception of the channels transmitted in the three modes for MIMO processing. CUT: channel-under-test, AWG: arrayed waveguide grating multipler, DAC: digital-to-analog converter, WSS: wavelength-selective switch, LO: local oscillator.

(b) Measured bit-error-rates (markers) and optical spectrum of all 96 channels after transmission over the 119km of few-mode-fibre with a mid-span amplifier.

To date, the longest transmission distance reported for MDM systems is 1200km, where a recirculating loop was utilized with a 30km span consisting of two FMF sections having DMGDs of opposite sign[33]. Mode multiplexers and demultiplexers were placed before and after the FMF span in the loop so that single-mode EDFAs could be used. Recently, a net capacity of more than 57Tb/s (after subtracting the overhead for FEC) was demonstrated in a 119km MDM system (Figure above(a)) consisting of WDM channels each at 200Gb/s. The FMF supported three spatial modes, and an inline MM-EDFA provided 18dB of gain per mode, making this the first WDM FMF system to utilize a mid-span MMF amplifier. The 12bit/s/Hz spectral efficiency and 57Tb/s capacity are the highest reported so far for MDM transmission. The highest number of spatial modes to be utilized as separate information channels was reported by Ryf and co-workers, who used six spatial modes (LP01, LP11, LP21, and LP02) and 12 x 12 MIMO.

Combining MCF and FMF Concepts

The multicore fibres with coupled cores allow increased core density and/or the cores' effective areas to be increased to minimize nonlinear effects. Due to the strong crosstalk between cores, the light propagation in the fibre can be described by the supermodes of the composite fibre structure. A single-channel transmission experiment with a 24-km homogenous three-core fibre with 104um effective areas was reported, where all six (space and polarization) modes were launched and jointly detected. The large crosstalk of about -4dB was almost completely suppressed by coherent 6x6 MIMO processing. In other work, an all-solid-glass microstructured fibre with three large (~129um) effective area cores at 29.4um pitch was utilized in several experiments. Using 6×6 MIMO processing, transmission over 1200km was achieved for a single 20-Gbaud-QPSK channel and over 4200km for five WDM channels, a record distance at that time, thus demonstrating the feasibility of MIMO interference cancellation for long-haul transmission.

In order to gain a more significant capacity increase it is possible to combine the MC and MDM approaches, and indeed the first experiments in this area are now beginning. For example, it is possible to produce an array of spatially isolated MM rather than SM cores, enabling MDM to be overlaid directly on to the MC approach. To date fibres capable of supporting up to 21 different spatial modes (7 cores each supporting 3 modes) have been reported although, as of yet, only relatively rudimentary system measurements have been performed. In a related approach, a hybrid MCF with 12 single-mode cores and two few-mode cores supporting three spatial modes each has been utilized to demonstrate transmission of 1.05 Pb/s capacity. Although the transmission distance was only 3km, the experiment was the first to achieve spectral efficiency beyond 100 bit/s/Hz. These approaches are extremely challenging from a component perspective, nevertheless they provide interesting opportunities both in terms of constraining DSP complexity and allowing very high spatial channel densities.

SDM Networking and Switching

Although to date the majority of the SDM demonstrations have consisted of point-to-point transmission, recent efforts are contemplating switching strategies and elements that could support flexible optical routing. Consensus is building that SDM networks should utilize spatial superchannels (i.e. groups of same-wavelength subchannels transmitted on separate spatial modes but routed together, where the spatial modes could be the regular modes in a MMF/FMF, super- modes in a strongly coupled multi-core fibre, or the fundamental modes of each individual single- mode core in an 'uncoupled' multi-core fibre). Such a strategy could provide sufficient granularity for efficient routing and facilitate ROADM integration, and could help to simplify network design since the modes are routed as one entity, foster transceiver integration (e.g. share a single source laser in the transmitter and a single local oscillator in the receiver), and lighten the DSP load by exploiting information about common-mode impairments such as dispersion and phase fluctuations.

As a first-step towards a FMF-compatible ROADM, an optical add-drop multiplexer (OADM) comprising two cascaded, free-space, thin-film filters has been demonstrated for the two orthogonal LP11 modes. Recently, Amaya reported switching in the space, frequency and time dimensions in an elastic SDM and multi-granular network that included two 7-core MCF links. Space switching was achieved via an optical backplane that interconnected MCF/SMF fibre inputs, functional modules, and MCF/SMF fibre outputs; however, there was a high degree of complexity due to the required demultiplexing of the signals transmitted on the various cores MCF at each node. Future work is needed to examine the trade-offs between the switching granularity and resulting complexity in SDM networks.

Polarization-division Multiplexing

A physical layer technique for multiplexing signals carried by magnetism waves, permitting 2 channels of data to be transfer on constant carrier frequency by victimization waves of 2 orthogonal polarization states. it's utilized in microwave links like television system downlinks to increase the information carry by victimization 2 orthogonally polarized feed antennas in satellite dishes. it's conjointly utilized in fiber optic communication by transmittal separate left and right circularly polarized light-weight beams through constant glass fiber. Polarization techniques has been long been utilized in radio transmission by scale back interference between channels, significantly at VH frequencies and beyond.

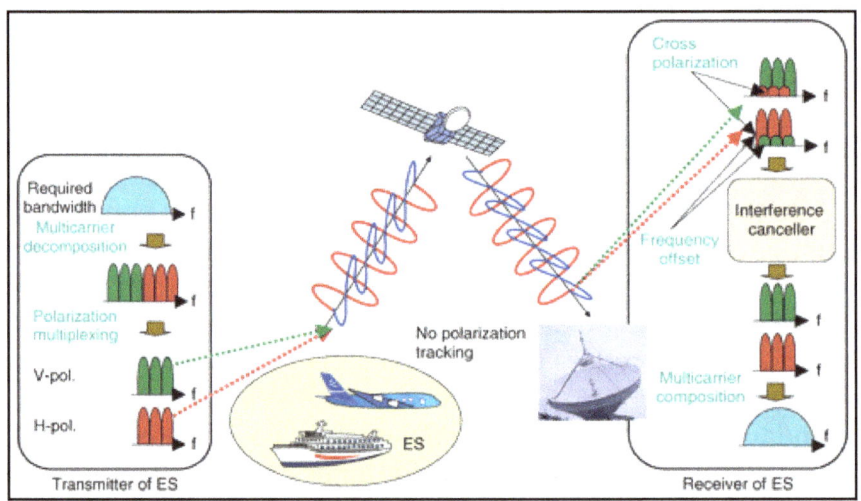

Under some circumstances, the info-rate of a link is enhanced by transmittal 2 separate mediums of radio waves on particular frequency, victimization orthogonal polarization. As an example, in purpose to purpose microwave links, the transmittal antenna will has 2 feed antennas; a vertical feed antenna that used to transmits microwaves with their field of force vertical (vertical polarization), and a horizontal feed antenna that used to transmits microwaves on constant frequency with their field to force horizontally (horizontal polarization). These separate vertical and horizontal feed antennas at the receiver end. For space and satellite communication, orthogonal circular polarization is preferred touse instead, (i.e. right- and left-handed), reason is the sense of circular polarization isn't enhanced by the relative orientation of the antenna in area.

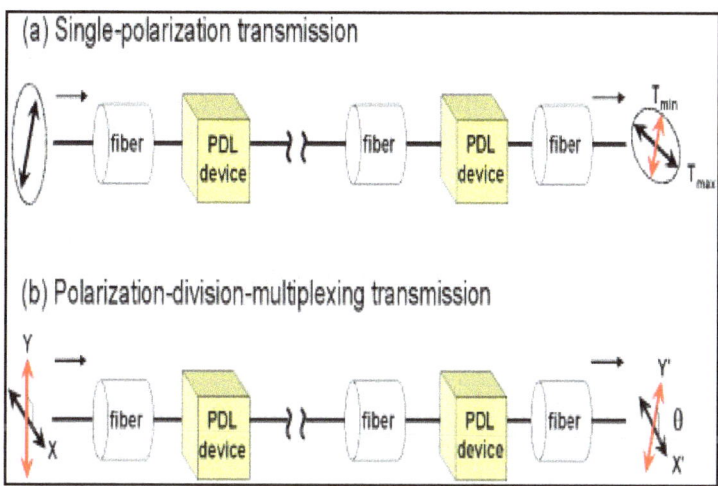

(a) Single-polarization transmission

(b) Polarization-division-multiplexing transmission

Radio

Polarization techniques have long been used in radio transmission to reduce interference between channels, particularly at VHF frequencies and beyond.

Under some circumstances, the data rate of a radio link can be doubled by transmitting two separate channels of radio waves on the same frequency, using orthogonal polarization. For example, in point to point terrestrial microwave links, the transmitting antenna can have two feed antennas; a vertical feed antenna which transmits microwaves with their electric field vertical (vertical polarization), and a horizontal feed antenna which transmits microwaves on the same frequency with their electric field horizontal (horizontal polarization). These two separate channels can be received by vertical and horizontal feed antennas at the receiving station. For satellite communications, orthogonal circular polarization is often used instead, (i.e. right- and left-handed), as the sense of circular polarization is not changed by the relative orientation of the antenna in space.

A dual polarization system comprises usually two independent transmitters, each of which can be connected by means of waveguide or TEM lines (such as coaxial cables or stripline or quasi-TEM such as microstrip) to a single-polarization antenna for its standard operation. Although two separate single-polarization antennas can be used for PDM (or two adjacent feeds in a reflector antenna), radiating two independent polarization states can be often easily achieved by means of a single dual-polarization antenna.

When the transmitter has a waveguide interface, typically rectangular in order to be in single-mode regione at the operating frequency, a dual-polarized antenna with a circular (or square) waveguide port is the radiating element chosen for modern communication systems. The circular or square waveguide port is needed so that at least two degenerate modes are supported. An ad-hoc component must be therefore introduced in such situations to merge two separate single-polarized signals into one dual-polarized physical interface, namely an ortho-mode transducer (OMT).

In case the transmitter has TEM or quasi-TEM output connections, instead, a dual-polarization antenna often presents separate connections (i.e. a printed square patch antenna with two feed points), and embeds the function of an OMT by means of intrinsically transferring the two excitation signals to the orthogonal polarization states.

A dual-polarized signal thus carries two independent data streams to a receiving antenna, which can itself be a single-polarized one, for receiving only one of the two streams at a time, or a dual-polarized model, again relaying its received signal to two single-polarization output connectors (via an OMT if in waveguide).

The ideal dual-polarization system lies its foundation onto the perfect orthogonality of the two polarization states, and any of the single-polarized interfaces at the receiver would theoretically contain only the signal meant to be transmitted by the desired polarization, thus introducing no interference and allowing the two data streams to be multiplexed and demultiplexed transparently without any degradation due to the coexistence with the other.

Companies working on commercial PDM technology include Siae Microelettronica, Huawei and Alcatel-Lucent.

Some types of outdoor microwave radios have integrated orthomode transducers and operate in both polarities from a single radio unit, performing cross-polarization interference cancellation (XPIC) within the radio unit itself. Alternatively, the orthomode transducer may be built into the antenna, and allow connection of separate radios, or separate ports of the same radio, to the antenna.

CableFree 2+0 XPIC Microwave Link showing OMT and two ODUs connected to H & V polarity ports.

Cross-polarization Interference Cancellation (XPIC)

Practical systems, however, suffer from non-ideal behaviors which mix the signals and the polarization states together:

- The OMT at the transmitting side has a finite cross-polarization discrimination (XPD) and thus leaks part of the signals meant to be transmitted in one polarization to the other.

- The transmitting antenna has a finite XPD and thus leaks part of its input polarizations to the other radiated polarization state.

- Propagation in presence of rain, snow, hail creates depolarization, as part of the two impinging polarizations is leaked to the other.

- The finite XPD of the receiving antenna acts similarly to the transmitting side and the relative alignment of the two antennas contributes to a loss of system XPD.

- The finite XPD of the receiving OMT likewise further mixes the signals from the dual-polarized port to the single-polarized ports.

As a consequence, the signal at one of the received single-polarization terminals actually contains a dominant quantity of the desired signal (meant to be transmitted onto one polarization) and a minor amount of undesired signal (meant to be transported by the other polarization), which represents an interference over the former. As a consequence, each received signal must be cleared of the interference level in order to reach the required signal-to-noise-and-interference ratio (SNIR) needed by the receiving stages, which may be of the order of more than 30 dB for high-level M-QAM schemes. Such operation is carried out by a cross-polarization-interference cancellation (XPIC), typically implemented as a baseband digital stage.

Compared to spatial multiplexing, received signals for a PMD system have a much more favourable carrier-to-interference ratio, as the amount of leakage is often much smaller than the useful signal, whereas spatial multiplexing operates with an amount of interference equal to the amount of useful signal. This observation, valid for a good PMD design, allows the adaptive XPIC to be designed in a simpler manner than a general MIMO cancelling scheme, since the starting point (without cancellation) is typically already sufficient for establishing a low-capacity link by means of a reduced modulation.

An XPIC typically acts on one of the received signals "C" containing the desired signal as dominant term and uses the other received "X" signal too (containing the interfering signal as dominant term). The XPIC algorithm multiplies the "X" by a complex coefficient and then adds it to the received "C". The complex recombination coefficient is adjusted adaptively to maximize the MMSE as measured on the recombination. Once the MMSE is improved to the required level, the two terminals can switch to high-order modulations.

Optical Polarization Division Multiplexing

Optical polarization division multiplexing is based on transmitting independently modulated signals over orthogonal polarizations of the same optical wavelength. In optics, the state of optical polarization (SOP) is defined as the shape traced out by the electric-field vector of the transmitted light in a fixed plane. figure(a) shows some examples of SOP, including linear, circular, and elliptical polarization states. Linear polarization is obtained when the direction of the electric vector is constant (the electric-field vectors along with and directions are in phase). When exactly ninety degrees out of phase, we have circular polarization. There are right-handed or left-handed circular/elliptical polarizations, depending on which direction the electric-field vector rotates, that is, if the electric-field vector is seen rotating clockwise or counterclockwise.

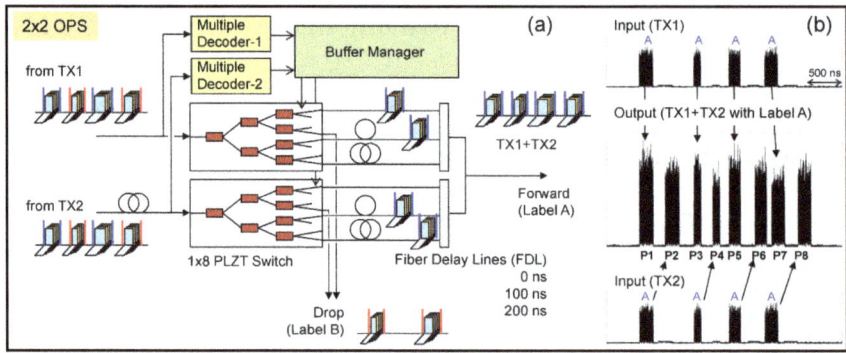

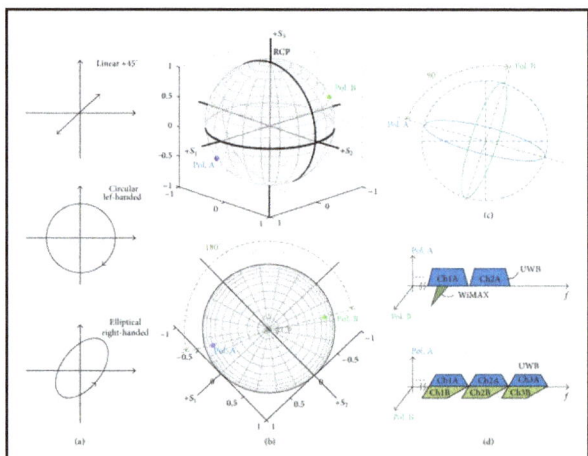

Figure: (a) Examples of linear, circular, and elliptical SOP. (b) Orthogonal polarization representation in Poincaré sphere. (c) Example of orthogonal ellipses of polarization. (d) Spectrum allocation examples for PDM transmission systems.

Figure above (b) shows an example of two orthogonal SOP represented in the Poincaré sphere. The polarization state is defined using the Stokes parameters that can be represented in the Poincaré sphere over the three-dimensional vector of Cartesian coordinates. The right circular polarization pole (RCP) is included in the sphere for reference. In the example presented in figures above (b) and (c), the polarization states were measured with Optellios PS2300 optical polarization analyzer. As it is described in figure (b), two polarizations are orthogonal if they are separated by 180° in the Poincaré sphere. Figure (c) shows the polarization ellipses for each polarization. In the representation of polarization ellipses, two states are orthogonal if they have a 90° phase shift (as sine and cosine).

The polarization multiplexing method can be explained from the similarity with the techniques used in microwave communications. In wireless links, the user bandwidth can be improved if two orthogonally polarized radio frequency signals are transmitted. At the receiver, two antennas with different polarization and orientation are used to discriminate each of the signals. The same occurs in optical systems: at the receiver, the two orthogonal states of polarization are detected, obtaining each of the modulated signals independently. If the orthogonality is maintained through the optical system, at the receiver, each of the modulated signals can be recovered. But SOP orthogonality is degraded due to the propagation in optical networks due to stress in the glass fiber (bending and twisting), moving the fiber, or even ambient temperature changes. For this reason, it is important to evaluate if the crosstalk due to cross-polarization interference is limiting the performance at the receiver.

Several techniques have been proposed to mitigate the interference coming from cross-polarization. One technique is based on frequency interleaving of the adjacent channels that are orthogonally polarized. Polarization interleave multiplexing systems have achieved experimentally up to 1.6 bit/s/Hz spectrum efficiency as demonstrated.

The main impairments suffered by optical PDM systems are due to cross-phase modulation (XPM) and polarization-mode dispersion (PMD). Nelson assessed the impairments caused by first-order PMD in the fiber using a non-return-to-zero (NRZ) polarization-multiplexed system with 40 Gb/s per channel and 0.8 bit/s/Hz spectral efficiency. The measured system penalty of 1 dB in instantaneous differential group delay (DGD) pointed out that polarization-multiplexed systems are five times more sensitive to PMD compared with nonpolarization-multiplexed systems due to crosstalk. However, a 121.9 Gb/s polarization-multiplexed transmission using coherent detection was demonstrated by Jansen with 4 OFDM channels at 2 bit/s/Hz spectral efficiency over 1000 km of standard-single mode fiber (SSMF).

The receiver of PDM-OFDM system can be seen as a multiple input multiple output (MIMO) system. Thus, digital signal processing (DSP) algorithm based on MIMO processing can be used for improving the optical transmission, as proposed in the literature for wireless communication. Recently, PDM systems have been demonstrated as a viable solution for fully standardized wireless MIMO provision using radio-over-fiber polarization-multiplexed long-reach optical transmission. Two different optical detection approaches can be implemented: direct or coherent detection. For simplicity of the receiver, in this paper, we will focus on direct detection, but the network reach could be further extended using coherent detection or adding MIMO digital processing. For example, a PON architecture of 20 km of SSMF was demonstrated with a 40 Gb/s PDM-OFDM transmission including MIMO digital signal processing at the receiver.

Orbital Angular Momentum Multiplexing

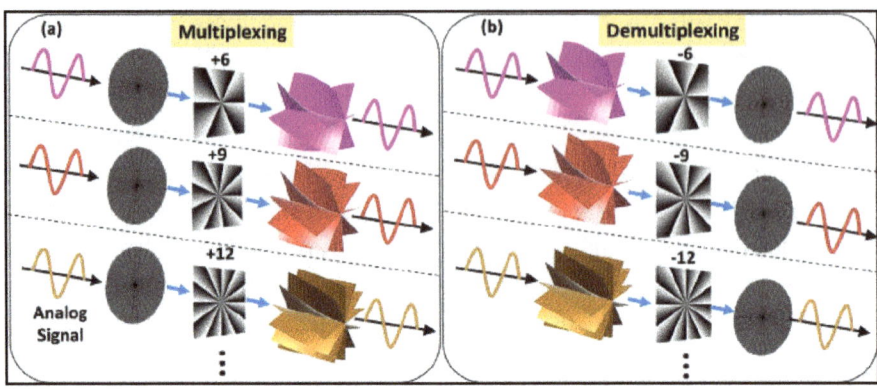

The orbital angular momentum (OAM) multiplexing method is an emerging wireless transmission technology that exploits a physical property of electromagnetic waves characterized by a helical phase front in the propagation direction. The propagating beams with a distinct number of phase rotations, that is, OAM modes, are orthogonal to one another and create multiple orthogonal channels. From a wireless communication perspective, the beauty of OAM multiplexing is that it generates multiple orthogonal channels in a line-of-sight channel environment without complex signal processing such as channel diagonalization.

Research on electromagnetic waves carrying OAM modes goes back a long way. Some work in this field dates from the early 1900s. In the 1980s, the multiplexing concept using OAM modes was suggested by researchers studying circular array antennas. From the late 1990s to the 2000s, similar concepts were studied in various fields, including radio astronomy, photonics, and free-space wireless communication. From the early 2010s, interest was drawn again to the wireless communication field, influenced by maturing technologies using mm-wave band communications.

Studies related to OAM multiplexing in the wireless communication field are categorized into those focusing on antenna design and beam generation, proof of concept experiments signal processing methods and system studies such as those concerning capacity analysis and link budget.

Much work has been done on antenna design and beam generation since the renewal of interest in the early 2010s. Various antenna designs using helicoidally deformed parabolic antennas, spiral phase plates (SPPs), holographic plates (HPs), and elaborately tuned planar SPPs have been reported. Despite some reports of successful transmissions, it seems to be difficult to perform multiplexing in a practical manner using such antennas since each OAM mode needs a differently located antenna. However, a uniform circular array (UCA) and multiple UCAs are considered to be suitable for OAM multiplexing because they can transmit coaxially aligned multiple streams simultaneously.

The feasibility of OAM multiplexing has been validated in many different experiments Yan. successfully demonstrated 32-Gbit/s OAM multiplexing over a 60-GHz mm-wave band with four concurrent OAM modes (Modes = −3, −1, 1, and 3) with 16-quadrature amplitude modulation (QAM). The transmission distance was 2.5 m, and four SPPs were used for multiplexing with a 4-to-1 combiner. Mahmouli conducted 4-Gbit/s uncompressed video transmission over a 60-GHz mm-wave band using HPs and SPPs. However, as of yet, no Gbit/s-level transmission experiments over 10 m have been reported.

OAm Beam Generation and Separation

To generate the beam carrying the OAM mode n (L = n), antenna elements are connected with phase shifters that make n × 360 degrees of rotation. Examples of beam generations of OAM modes 0, 1, and 2 using UCAs consisting of eight antenna elements are shown in figure above Note that it is possible to use either a single UCA or multiple UCAs for multiple OAM mode generation. In the former case, superposed beams are transmitted by a single UCA. In the latter case, concentric multiple UCAs are used.

The separation of beams carrying OAM modes can be done in a way similar to that for generation using antenna elements connected with phase shifters that make opposite rotation directions. As long as the number of antenna elements is larger than 2n, rotations of n × 360 degrees are orthogonal to one another. Therefore, each OAM mode can be separated from mixed OAM mode signals without aliasing. An example of each antenna element phase corresponding to the example above is shown in figure aboveSuch beam separation can also be done by using a single UCA or multiple UCAs as in the beam generation. Note that a divider is fitted between antenna elements and phase shifters in the former case.

Properties of OAM Beams

Some examples of intensity and phase distributions of beams carrying multiple OAM modes generated by a single UCA are shown in figure below. At the Rx antenna, the diffraction pattern generated by a single transmitting UCA is calculated by the summation of the electric field generated from each antenna element. Since the diffraction pattern of the UCA can be approximated as a Bessel beam, the electric field distribution of the beam carrying OAM mode L is often expressed by the Bessel beam's equation as below:

$$v_L(r,\theta,z) = \frac{\lambda \exp\left[(2\pi/\lambda)\right]\sqrt{r^2+z^2}}{4\pi\sqrt{r^2+z^2}} \cdot i^{-L} \exp\left[iL\theta\right] \cdot J_L$$

$$\left(\frac{2\pi rD}{\lambda\sqrt{r^2+z^2}}\right)$$

Where $J_L(\cdot)$, λ, and D respectively denote the Lth order Bessel function of the first kind, the wavelength of the carrier frequency, and the radius of the transmitting

UCA. Equation above is represented in cylindrical coordinates, where r and θ are respectively the radius and azimuthal angle at the Rx plane that is vertical to the beam propagation direction, and z is the distance between the centers of the transmitter (Tx) and Rx UCAs.

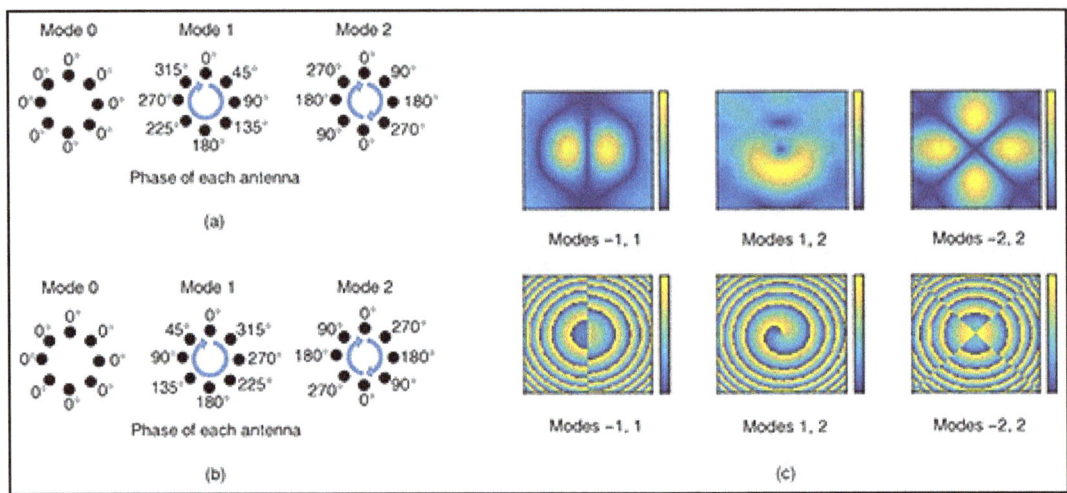

Figure: (a) Generation of OAM modes with a UCA, (b) separation of OAM modes with a UCA, and (c) intensity distribution (above) and phase distribution (below) of combined OAM modes.

Under the free-space propagation or AWGN (additive white Gaussian noise) channel, the intensity and phase of the received signals at a certain location can be analytically obtained by equation above with parameters consisting of the radius (D) of the Tx UCA, wavelength (λ), and the Lth order Bessel function of the first kind ($J_L(\cdot)$).

Challenges

This subsection highlights three major issues that have to be resolved in order to fully exploit the potential of OAM multiplexing. These issues are as follows:

(1) Beam divergence

Beams carrying OAM modes diverge along with their propagation as shown in figure. Other than OAM mode 0, the locations of the first Rx peak intensities of OAM modes diverge as the propagation distance increases. It may be considered that various beam-forming technologies can generate a sharp non-OAM carrying beam of which the divergence is not significant. However, the divergence of OAM carrying beams is determined by the Tx UCA radius and wavelength. It is therefore necessary to increase the physical size of the Tx UCA or to use a higher frequency in order to reduce the divergence of OAM carrying beams. Since both the Tx antenna size and frequency band are usually not tunable design factors, we leave this issue as an open problem while suggesting that the Tx antenna size be set as large as the physical environment allows.

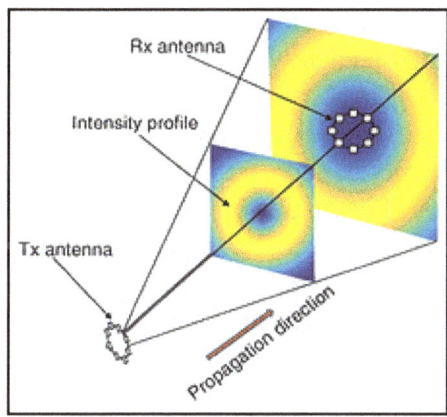

(2) Mode-dependent performance degradation

The beam that carries different OAM modes yields different locations of the peak intensities. This non-identical peak Rx intensity results in mode-dependent performance degradation in accordance with the location of the Rx antenna. In addition, the required Rx antenna size to capture the peak Rx power becomes larger as the number of OAM modes increases. If the Rx antenna size is limited to a certain size, some OAM modes might not have the peak Rx power. Correspondingly, the performance degradation for higher OAM modes becomes more severe.

Proposed Rx Antenna Design and OAM Beam Separation

1. Rx antenna design

In view of the distinct beam propagation and divergence among OAM modes, it may be a good approach to put the Rx antenna for each OAM mode at its optimum or near-optimum location. For example, the use of many concentric Rx UCAs that capture OAM modes at their high Rx SNR region may lessen the beam divergence problem and mode-dependent performance degradation. In such concentric Rx UCA cases, the outer UCA is generally used for the reception of a higher OAM mode. Since the UCA antenna radius increases by the power of two in such cases, a spacious antenna is necessary to capture the higher OAM mode signals in this approach.

To provide a practical solution that allows higher OAM mode signals to be received at a high Rx SNR while maintaining a reasonable antenna size, we present a simple but practical Rx antenna design and corresponding beam separation method. Our idea is based on the fact that there are specific location sets of which phase differences are 90 or 180 degrees. Since such specific location sets depend not on the Euclidean distance but on the angle conditions that will be explained later, these specific location sets are invariant in terms of the distance between Tx and Rx. This provides the flexibility in the system design.

The concept of the proposed Rx antenna for concurrent transmission of seven OAM modes, including OAM modes −3, −2, −1, 0, 1, 2, and 3, is illustrated in figue below. We use a four-antenna-element set to receive a pair of OAM modes of which the absolute values are identical and the signs differ from each other, for example, OAM modes 1 and −1. We also use an antenna at the center for OAM mode 0. In figure, the outermost, middle, and innermost four-antenna-element sets are respectively for OAM modes 3 and −3, 2 and −2, and 1 and −1. Four antenna elements in each set are located equidistant from the center and form an 'X' type configuration. The angles of the two upper antenna elements of each set are respectively 30, 45, and 90 degrees. The angles of the two lower antenna elements of each set are the same as those of the upper ones. Note that such angles become narrower as the number of OAM modes increases. Therefore, in contrast to the UCA case, the necessary area to capture the higher OAM modes does not increase by the power of two in its radius.

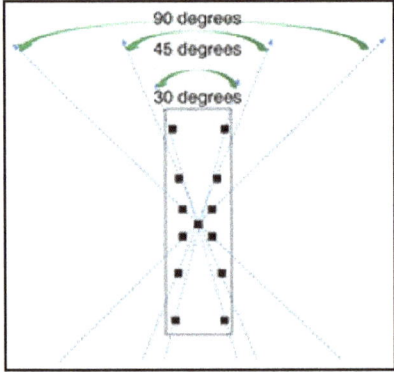

Fig: Proposed Rx antenna.

2. OAM beam separation

With the presented Rx antenna configuration, OAM beams are separated by analog cancellation followed by digital cancellation. We explain the details on the beam separation, which consist of the following three steps, referring to figure below:

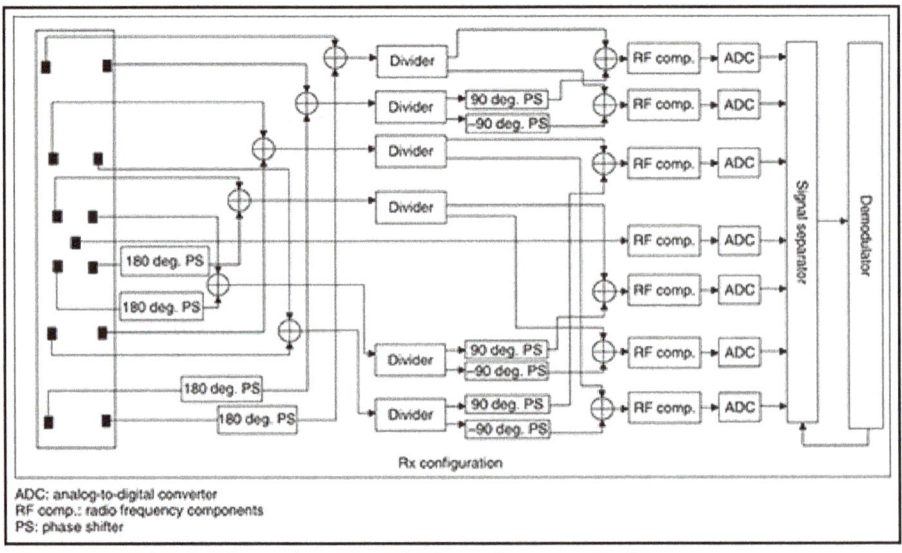

Fig: Configuration of Rx device using proposed Rx antenna.

Step 1. Analog separation of odd and even OAM modes

Diagonally located antenna elements in each four-antenna-element set are combined by either an equal-phase or a reverse-phase combination using 2-to-1 analog combiners. Before these combinations, in the four-antenna-element sets installed for odd OAM modes, the two antenna elements at the bottom are first inserted into the 180-degree phase shifters as shown in figure above. With these equal-phase or reverse-phase combinations, the combined outputs can only bear either an odd or even OAM mode.

Step 2. Analog extraction of each OAM mode

In this step, we begin by explaining an example of the extraction of OAM modes −3 and 3. A conceptual example to extract OAM mode 3 is shown in figure below. Two reverse-phase combined outputs from the outermost four-antenna-element set contain only odd OAM modes (i.e., OAM mode −3, −1, 1, and 3) while even OAM mode signals are canceled out.

(a) One output of the reverse-phase combination using two antenna elements among the outermost four-antenna-element set;

(b) Output of 90-degree phase shifter using another output of the reverse-phase combination using remaining antenna elements among the outermost four-antenna-element set;

(c) Output of −90-degree phase shifter using another output of the reverse-phase combination using remaining antenna elements among the outermost four-antenna-element set;

(d) Combined signal of (b) and (c) (OAM mode 3 is extracted).

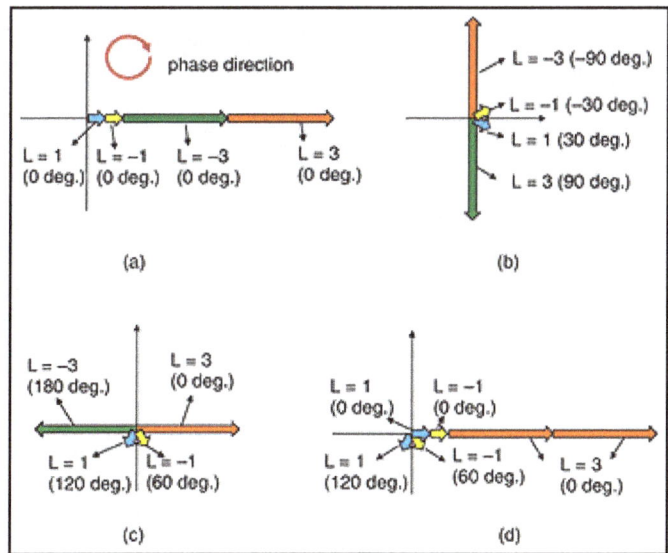

Fig: Conceptual example of OAM signal separation (OAM mode 3).

Each of the two combined outputs is again divided into two signals by a 1-to-2 divider. One output of the two dividers is inserted into the 90 and −90 degree phase shifters. The outputs of these phase shifters are respectively combined again with the outputs of the other two dividers. These combined signals respectively yield OAM modes −3 and 3 containing some interference from OAM modes −1 and 1. The interference is resolved by digital processing in step 3. OAM modes −1 and 1, and −2 and 2 are similarly extracted using two equal-phase combined outputs from the middle and innermost four-antenna-element sets. Extraction of OAM mode 0 is directly obtained from the antenna element at the center since all OAM modes disappear other than OAM mode 0 at the center.

Step 3. Digital pruning of each OAM mode

Each of the OAM mode signals extracted in the previous step is further pruned by digital signal processing such as successive interference cancellation or multiple-input multiple-output equalization. This step can be skipped when the residual interference after step 2 is negligible, or it can be intentionally skipped to reduce the complexity.

Evaluation: Proof of Concept Experiments

We conducted proof of concept experiments to examine the feasibility of OAM multiplexing. First, beams carrying OAM modes were generated using an unmodulated signal at 5.2 GHz. Their propagation was also investigated. Second, wireless communication experiments using quadrature phase shift keying (QPSK) and 16-QAM modulated signals at the same frequency band were conducted using our proposed antenna. Our experimental environment, Tx antenna and its radiation pattern, and the proposed Rx antenna are shown in figure:

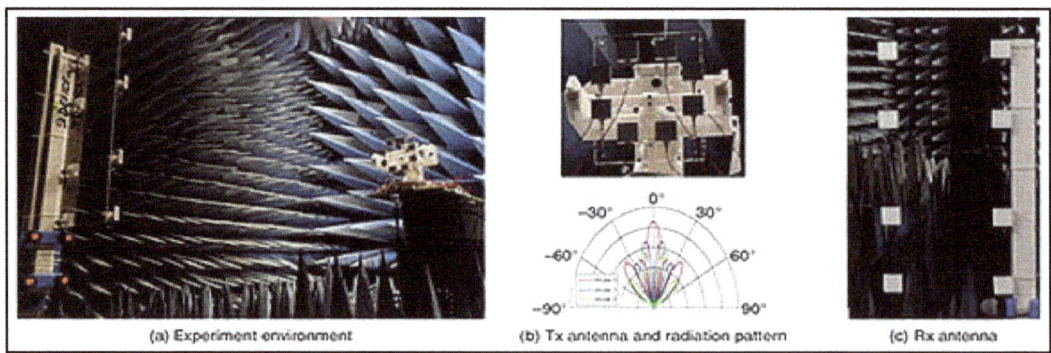

(a) Experiment environment   (b) Tx antenna and radiation pattern   (c) Rx antenna

Figure: Experimental setups.

OAM Beam Generation and Propagation Experiments

Three beams respectively carrying OAM modes 0, 1, and 2 were generated using a single Tx UCA at different times. The results of measuring intensity and phase distributions are shown in figure(a). The figure shows the average intensity distribution normalized to the peak intensity of OAM mode 0 and the phase distribution in terms of the distance from the Rx antenna center. We confirmed from the measurement results that the experimental generated OAM beam propagation patterns were in good agreement with theory.

Precisely, an unmodulated signal at 5.2 GHz was generated and fed into a 1-to-8 divider. The divider's eight outputs were then respectively connected to eight tunable phase shifters. Each tunable phase shifter was connected to the antenna elements of the Tx UCA. We set the phases of the tunable phase shifters in order to independently generate beams carrying OAM modes. For example, we set the phases from 0 degrees to 315 degrees by increasing them in 45-degree increments to generate OAM mode 1. This yielded 360 degrees of phase rotation. To capture the intensity and phase distributions at the Rx plane, we used a horn Rx antenna and a moving positioner. While transmitting the generated OAM beam, we recorded the intensity and phase distribution at two-dimensional grids of the Rx plane.

Our experimental setup in the shielding room is shown in figure(a), and the Tx UCA (top) and its radiation pattern (bottom) are in figure(b). Note that we conducted these experiments using a horn Rx antenna and a moving positioner. The distance between the Tx UCA and the Rx horn antenna was around 235 cm (40.7 λ), and the diameter of the Tx UCA was around 11.54 cm (2 λ). The positioner was moved over a 21 × 21 square grid. One span of the grid was 5.77 cm (1 λ), and the measured area at the Rx plane covered around 115.38 cm × 115.38 cm. The center of the grid was set to be the peak intensity spot for OAM mode 0, while it was set to be the null points for OAM modes 1 and 2.

Modulated Signal Transmission using Oam Beams

Here, we describe the results obtained in a wireless communication experiment using modulated signals. Except for the Rx antenna, the experimental environment was the same as that used in previous experiments. Instead of using a horn antenna, we used our proposed antenna shown in figure(c). The width and height of the Rx antenna were respectively 29 cm and 70 cm. QPSK and 16-QAM modulations were used for both uncoded and coded (1/2 rate low-density parity check (LDPC)) cases. Orthogonal frequency division multiplexing was carried out with 64 subcarriers over a 20-MHz signal bandwidth. Of the 64 subcarriers, 16 were used. Due to the practical limitations of the experimental setup, this experiment was conducted using a single stream while varying the OAM modes among −2, −1, 0, 1, and 2. The OAM multiplexing was evaluated by combining these single stream signals by offline processing.

The constellation maps of QPSK and 16-QAM signals obtained from a single stream are shown respectively in figure below (b) and (c). Their error vector magnitude values were respectively 14.18% and 13.5% with OAM mode 1, and 14.3% and 15.65% with OAM mode 2. Through off-line combining of the received signals of three single streams (OAM modes 0, 1, and 2) that were obtained by analog extraction, we confirmed that a bit error rate less than 0.001 is feasible with LDPC coding (rate 1/2). More enhanced performance is expected when digital pruning is further applied. These results validate the feasibility of wireless transmission by OAM multiplexing.

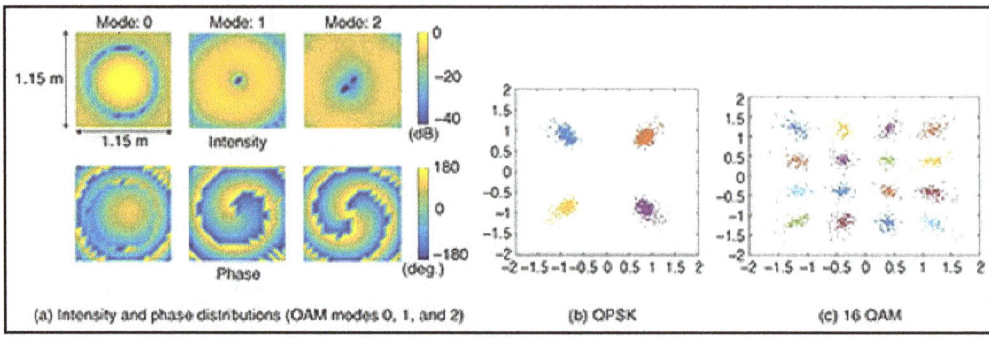

Figure: Experimental results.

Wavelength division Multiplexing

Wavelength division multiplexing (WDM) is a technology or technique modulating numerous data streams, i.e. optical carrier signals of varying wavelengths (colors) of laser light, onto a single optical fiber. WDM enables bi-directional communication as well as multiplication of signal capacity.

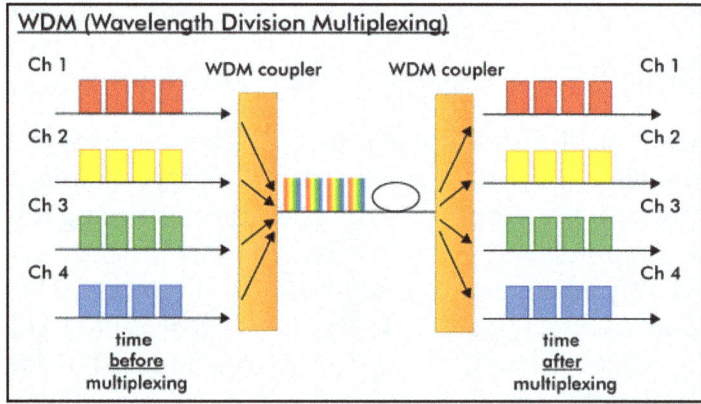

WDM is actually frequency division multiplexing (FDM) but referencing the wavelength of light as opposed to the frequency of light. However, since wavelength and frequency have an inverse relationship (shorter wavelength means higher frequency), the WDM and FDM terms actually describe the same technology – light in optical cable used to carry data and communication signals.

Wavelength division multiplexing systems can combine signals with multiplexing and split them apart with a demultiplexer. And with the proper fiber cable, the two can be done simultaneously; moreover, these two devices can also function as an add/drop multiplexer (ADM), i.e. simultaneously adding light beams while dropping other light beams and rerouting them to other destinations and devices. Formerly, such filtering of light beams was done with etalons, devices called Fabry–Pérot interferometers using thin-film-coated optical glass. The first WDM technology was conceptualized in the early 1970s and realized in the laboratory in the late 1970s; but these only combined two signals, and many years later were still very expensive.

As of 2011, WDM systems can handle 160 signals, which will expand a 10 Gbit/second system with a single fiber optic pair of conductors to more than 1.6 Tbit/second (i.e. 1,600 Gbit/s).

Typical WDM systems use single-mode optical fiber (SMF); this is optical fiber for only a single ray of light and having a core diameter of 9 millionths of a meter (9 μm). Other systems with multi-mode fiber cables (MM Fiber; also called premises cables) have core diameters of about 50 μm. Standardization and extensive research have brought down system costs significantly.

WDM systems are divided according to wavelength categories, generally course WDM (CWDM) and dense WDM (DWDM). CWDM operates with 8 channels (i.e., 8 fiber optic cables) in what is known as the "C-Band" or "erbium window" with wavelengths about 1550 nm (nanometers or billionths of a meter, i.e. 1550×10^{-9} meters). DWDM also operates in the C-Band but with 40 channels at 100 GHz spacing or 80 channels at 50 GHz spacing. Even newer technology, called Raman amplification, is using light in the L-Band (1565 nm to 1625 nm), approximately doubling these capacities.

Coarse WDM

Series of SFP+ transceivers for 10 Gbit/s WDM communications.

Originally, the term *coarse wavelength division multiplexing* (CWDM) was fairly generic, and meant a number of different things. In general, these things shared the fact that the choice of channel spacings and frequency stability was such that erbium doped fiber amplifiers (EDFAs) could not be utilized. Prior to the relatively recent ITU standardization of the term, one common meaning for coarse WDM meant two (or possibly more) signals multiplexed onto a single fiber, where one signal was in the 1550 nm band, and the other in the 1310 nm band.

In 2002 the ITU standardized a channel spacing grid for use with CWDM (ITU-T

G.694.2), using the wavelengths from 1270 nm through 1610 nm with a channel spacing of 20 nm. (G.694.2 was revised in 2003 to shift the actual channel centers by 1 nm, so that strictly speaking the center wavelengths are 1271 to 1611 nm). Many CWDM wavelengths below 1470 nm are considered "unusable" on older G.652 specification fibers, due to the increased attenuation in the 1270–1470 nm bands. Newer fibers which conform to the G.652.C and G.652.D standards, such as Corning SMF-28e and Samsung Widepass nearly eliminate the "water peak" attenuation peak and allow for full operation of all 18 ITU CWDM channels in metropolitan networks.

The 10GBASE-LX4 10 Gbit/s physical layer standard is an example of a CWDM system in which four wavelengths near 1310 nm, each carrying a 3.125 gigabit-per-second (Gbit/s) data stream, are used to carry 10 Gbit/s of aggregate data.

The main characteristic of the recent ITU CWDM standard is that the signals are not spaced appropriately for amplification by EDFAs. This therefore limits the total CWDM optical span to somewhere near 60 km for a 2.5 Gbit/s signal, which is suitable for use in metropolitan applications. The relaxed optical frequency stabilization requirements allow the associated costs of CWDM to approach those of non-WDM optical components.

CWDM is also being used in cable television networks, where different wavelengths are used for the *downstream* and *upstream* signals. In these systems, the wavelengths used are often widely separated, for example the downstream signal might be at 1310 nm while the upstream signal is at 1550 nm.

An interesting and relatively recent development relating coarse WDM is the creation of GBIC and small form factor pluggable (SFP) transceivers utilizing standardized CWDM wavelengths. GBIC and SFP optics allow for something very close to a seamless upgrade in even legacy systems that support SFP interfaces. Thus, a legacy switch system can be easily "converted" to allow wavelength multiplexed transport over a fiber simply by judicious choice of transceiver wavelengths, combined with an inexpensive passive optical multiplexing device.

Passive CWDM is an implementation of CWDM that uses no electrical power. It separates the wavelengths using passive optical components such as bandpass filters and prisms. Many manufacturers are promoting passive CWDM to deploy fiber to the home.

Dense WDM

Dense wavelength division multiplexing (DWDM) refers originally to optical signals multiplexed within the 1550 nm band so as to leverage the capabilities (and cost) of erbium doped fiber amplifiers (EDFAs), which are effective for wavelengths between approximately 1525–1565 nm (C band), or 1570–1610 nm (L band). EDFAs were originally developed to replace SONET/SDH optical-electrical-optical (OEO) regenerators, which they have made practically obsolete. EDFAs can amplify any optical signal in their operating range, regardless of the modulated bit rate. In terms of multi-wavelength signals, so long as the EDFA has enough pump energy available to it, it can amplify as many optical signals as can be multiplexed into its amplification band (though signal densities are limited by choice of modulation format). EDFAs therefore allow a single-channel optical link to be upgraded in bit rate by replacing only equipment at the ends of the link, while retaining the existing EDFA or series of EDFAs through a long haul route. Furthermore, single-wavelength links using EDFAs can similarly be upgraded to WDM links at reasonable cost. The EDFA's cost is thus leveraged across as many channels as can be multiplexed into the 1550 nm band.

DWDm Systems

At this stage, a basic DWDM system contains several main components:

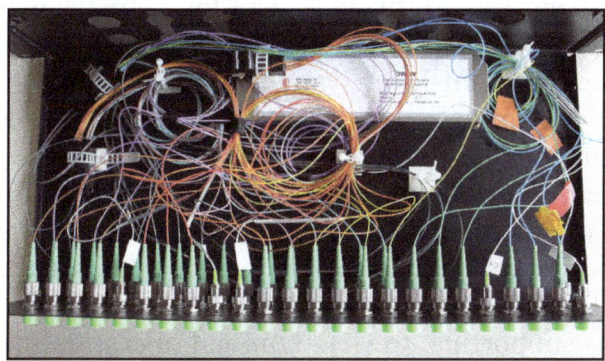

WDM multiplexer for DWDM communications.

1. A DWDM terminal multiplexer: The terminal multiplexer contains a wavelength-converting transponder for each data signal, an optical multiplexer and where necessary an optical amplifier (EDFA). Each wavelength-converting transponder receives an optical data signal from the client-layer, such as Synchronous optical networking [SONET /SDH] or another type of data signal, converts this signal into the electrical domain and re-transmits the signal at a specific wavelength using a 1,550 nm band laser. These data signals are then combined together into a multi-wavelength optical signal using an optical multiplexer, for transmission over a single fiber (e.g., SMF-28 fiber). The terminal multiplexer may or may not also include a local transmit EDFA for power amplification of the multi-wavelength optical signal. In the mid-1990s DWDM systems contained 4 or 8 wavelength-converting transponders; by 2000 or so, commercial systems capable of carrying 128 signals were available.

2. An intermediate line repeater: This is placed approximately every 80–100 km to compensate for the loss of optical power as the signal travels along the fiber. The 'multi-wavelength optical signal' is amplified by an EDFA, which usually consists of several amplifier stages.

3. An intermediate optical terminal, or optical add-drop multiplexer: This is a

remote amplification site that amplifies the multi-wavelength signal that may have traversed up to 140 km or more before reaching the remote site. Optical diagnostics and telemetry are often extracted or inserted at such a site, to allow for localization of any fiber breaks or signal impairments. In more sophisticated systems (which are no longer point-to-point), several signals out of the multi-wavelength optical signal may be removed and dropped locally.

4. A DWDM terminal demultiplexer: At the remote site, the terminal de-multiplexer consisting of an optical de-multiplexer and one or more wavelength-converting transponders separates the multi-wavelength optical signal back into individual data signals and outputs them on separate fibers for client-layer systems (such as SONET/SDH). Originally, this de-multiplexing was performed entirely passively, except for some telemetry, as most SONET systems can receive 1,550 nm signals. However, in order to allow for transmission to remote client-layer systems (and to allow for digital domain signal integrity determination) such de-multiplexed signals are usually sent to O/E/O output transponders prior to being relayed to their client-layer systems. Often, the functionality of output transponder has been integrated into that of input transponder, so that most commercial systems have transponders that support bi-directional interfaces on both their 1,550 nm (i.e., internal) side, and external (i.e., client-facing) side. Transponders in some systems supporting 40 GHz nominal operation may also perform forward error correction (FEC) via digital wrapper technology, as described in the ITU-T G.709 standard.

5. Optical Supervisory Channel (OSC): This is data channel which uses an additional wavelength usually outside the EDFA amplification band (at 1,510 nm, 1,620 nm, 1,310 nm or another proprietary wavelength). The OSC carries information about the multi-wavelength optical signal as well as remote conditions at the optical terminal or EDFA site. It is also normally used for remote software upgrades and user (i.e., network operator) Network Management information. It is the multi-wavelength analogue to SONET's DCC (or supervisory channel). ITU standards suggest that the OSC should utilize an OC-3 signal structure, though some vendors have opted to use 100 megabit Ethernet or another signal format. Unlike the 1550 nm multi-wavelength signal containing client data, the OSC is always terminated at intermediate amplifier sites, where it receives local information before re-transmission.

The introduction of the ITU-T G.694.1 frequency grid in 2002 has made it easier to integrate WDM with older but more standard SONET/SDH systems. WDM wavelengths are positioned in a grid having exactly 100 GHz (about 0.8 nm) spacing in optical frequency, with a reference frequency fixed at 193.10 THz (1,552.52 nm). The main grid is placed inside the optical fiber amplifier bandwidth, but can be extended to wider bandwidths. The first commercial deployment of DWDM was made by Ciena

Corporation on the Sprint network in June 1996. Today's DWDM systems use 50 GHz or even 25 GHz channel spacing for up to 160 channel operation.

DWDM systems have to maintain more stable wavelength or frequency than those needed for CWDM because of the closer spacing of the wavelengths. Precision temperature control of laser transmitter is required in DWDM systems to prevent "drift" off a very narrow frequency window of the order of a few GHz. In addition, since DWDM provides greater maximum capacity it tends to be used at a higher level in the communications hierarchy than CWDM, for example on the Internet backbone and is therefore associated with higher modulation rates, thus creating a smaller market for DWDM devices with very high performance. These factors of smaller volume and higher performance result in DWDM systems typically being more expensive than CWDM.

Recent innovations in DWDM transport systems include pluggable and software-tunable transceiver modules capable of operating on 40 or 80 channels. This dramatically reduces the need for discrete spare pluggable modules, when a handful of pluggable devices can handle the full range of wavelengths.

Wavelength-converting Transponders

At this stage, some details concerning wavelength-converting transponders should be discussed, as this will clarify the role played by current DWDM technology as an additional optical transport layer. It will also serve to outline the evolution of such systems over the last 10 or so years.

As stated above, wavelength-converting transponders served originally to translate the transmit wavelength of a client-layer signal into one of the DWDM system's internal wavelengths in the 1,550 nm band (note that even external wavelengths in the 1,550 nm will most likely need to be translated, as they will almost certainly not have the required frequency stability tolerances nor will it have the optical power necessary for the system's EDFA).

In the mid-1990s, however, wavelength converting transponders rapidly took on the additional function of signal regeneration. Signal regeneration in transponders quickly evolved through 1R to 2R to 3R and into overhead-monitoring multi-bitrate 3R regenerators:

1R

Retransmission basically, early transponders were "garbage in garbage out" in that their output was nearly an analogue "copy" of the received optical signal, with little signal cleanup occurring. This limited the reach of early DWDM systems because the signal had to be handed off to a client-layer receiver (likely from a different vendor) before the signal deteriorated too far. Signal monitoring was basically confined to optical domain parameters such as received power.

2R

Re-time and re-transmit transponders of this type were not very common and utilized a qua-si-digital Schmitt-triggering method for signal clean-up. Some rudimentary signal-quality monitoring was done by such transmitters that basically looked at analogue parameters.

3R

Re-time, re-transmit, re-shape 3R Transponders were fully digital and normally able to view SONET/SDH section layer overhead bytes such as A1 and A2 to determine signal quality health. Many systems will offer 2.5 Gbit/s transponders, which will normally mean the transponder is able to perform 3R regeneration on OC-3/12/48 signals, and possibly gigabit Ethernet, and reporting on signal health by monitoring SONET/SDH section layer overhead bytes. Many transponders will be able to perform full multi-rate 3R in both directions. Some vendors offer 10 Gbit/s transponders, which will perform Section layer overhead monitoring to all rates up to and including OC-192.

Muxponder

The muxponder (from multiplexed transponder) has different names depending on vendor. It essentially performs some relatively simple time-division multiplexing of lower-rate signals into a higher-rate carrier within the system (a common example is the ability to accept 4 OC-48s and then output a single OC-192 in the 1,550 nm band). More recent muxponder designs have absorbed more and more TDM functionality, in some cases obviating the need for traditional SONET/SDH transport equipment.

Reconfigurable Optical Add-drop Multiplexer (ROADM)

As mentioned above, intermediate optical amplification sites in DWDM systems may allow for the dropping and adding of certain wavelength channels. In most systems deployed as of August 2006 this is done infrequently, because adding or dropping wavelengths requires manually inserting or replacing wavelength-selective cards. This is costly, and in some systems requires that all active traffic be removed from the DWDM system, because inserting or removing the wavelength-specific cards interrupts the multi-wavelength optical signal.

With a ROADM, network operators can remotely reconfigure the multiplexer by sending soft commands. The architecture of the ROADM is such that dropping or adding wavelengths does not interrupt the "pass-through" channels. Numerous technological approaches are utilized for various commercial ROADMs, the tradeoff being between cost, optical power, and flexibility.

Optical Cross Connects (OXCs)

When the network topology is a mesh, where nodes are interconnected by fibers to form an arbitrary graph, an additional fiber interconnection device is needed to route the signals from an input port to the desired output port. These devices are called optical cross connectors (OXCs). Various categories of OXCs include electronic ("opaque"), optical ("transparent"), and wavelength selective devices.

Enhanced WDM

Cisco's Enhanced WDM system combines 1 Gb Coarse Wave Division Multiplexing (CWDM) connections using SFPs and GBICs with 10 Gb Dense Wave Division Multiplexing (DWDM) connections using XENPAK, X2 or XFP DWDM modules. These DWDM connections can either be passive or boosted to allow a longer range for the connection. In addition to this, CFP modules deliver 100 Gbit/s Ethernet suitable for high speed Internet backbone connections.

Shortwave WDM

Shortwave WDM uses vertical-cavity surface-emitting laser (VCSEL) transceivers with four wavelengths in the 846 to 953 nm range over single OM5 fiber, or 2-fiber connectivity for OM3/OM4 fiber.

Code division Multiplexing

Code Division Multiple Access (CDMA) is a sort of multiplexing that facilitates various signals to occupy a single transmission channel. It optimizes the use of available bandwidth. The technology is commonly used in ultra-high-frequency (UHF) cellular telephone systems, bands ranging between the 800-MHz and 1.9-GHz.

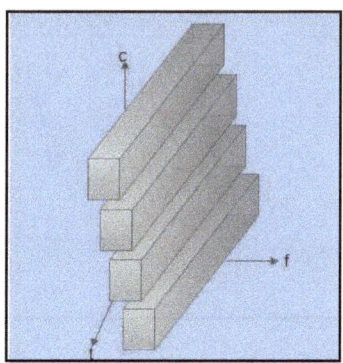

Code Division Multiple Access (CDMA).

Code Division Multiple Access system is very different from time and frequency multiplexing. In this system, a user has access to the whole bandwidth for the entire duration. The basic principle is that different CDMA codes are used to distinguish among the different users.

Techniques generally used are direct sequence spread spectrum modulation (DS-CD-MA), frequency hopping or mixed CDMA detection (JDCDMA). Here, a signal is generated which extends over a wide bandwidth. A code called spreading code is used to perform this action. Using a group of codes, which are orthogonal to each other, it is possible to select a signal with a given code in the presence of many other signals with different orthogonal codes.

Working of CDMA

CDMA allows up to 61 concurrent users in a 1.2288 MHz channel by processing each voice packet with two PN codes. There are 64 Walsh codes available to differentiate between calls and theoretical limits. Operational limits and quality issues will reduce the maximum number of calls somewhat lower than this value.

In fact, many different "signals" baseband with different spreading codes can be modulated on the same carrier to allow many different users to be supported. Using different orthogonal codes, interference between the signals is minimal. Conversely, when signals are received from several mobile stations, the base station is capable of isolating each as they have different orthogonal spreading codes.

The following figure shows the technicality of the CDMA system. During the propagation, we mixed the signals of all users, but by that you use the same code as the code that was used at the time of sending the receiving side. You can take out only the signal of each user.

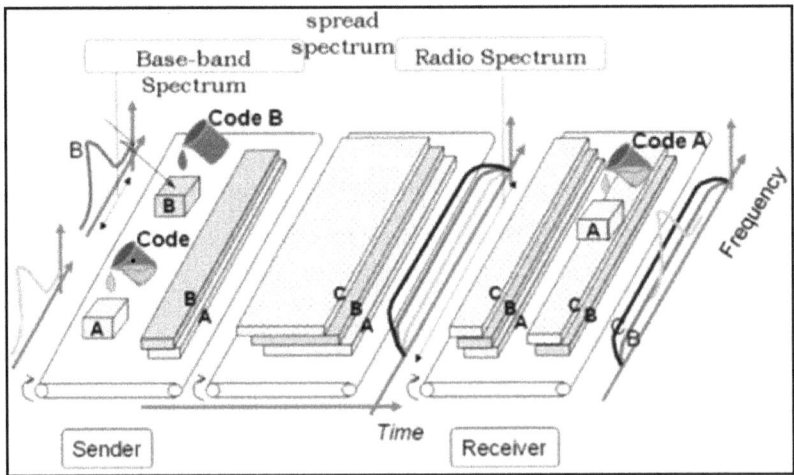

CDMA Capacity

The factors deciding the CDMA capacity are:

- Processing Gain.

- Signal to Noise Ratio.

- Voice Activity Factor.

- Frequency Reuse Efficiency.

Capacity in CDMA is soft, CDMA has all users on each frequency and users are separated by code. This means, CDMA operates in the presence of noise and interference.

In addition, neighboring cells use the same frequencies, which means no re-use. So, CDMA capacity calculations should be very simple. No code channel in a cell, multiplied by no cell. But it is not that simple. Although not available code channels are 64, it may not be possible to use a single time, since the CDMA frequency is the same.

Centralized Methods

- The band used in CDMA is 824 MHz to 894 MHz (50 MHz + 20 MHz separation).

- Frequency channel is divided into code channels.

- 1.25 MHz of FDMA channel is divided into 64 code channels.

Processing Gain

CDMA is a spread spectrum technique. Each data bit is spread by a code sequence. This means, energy per bit is also increased. This means that we get a gain of this.

P (gain) = 10log (W/R)

W is Spread Rate

R is Data Rate

For CDMA P (gain) = 10 log (1228800/9600) = 21dB

This is a gain factor and the actual data propagation rate. On an average, a typical transmission condition requires a signal to the noise ratio of 7 dB for the adequate quality of voice.

Translated into a ratio, signal must be five times stronger than noise,

Actual processing gain = P (gain) - SNR

= 21 − 7 = 14dB

CDMA uses variable rate coder,

The Voice Activity Factor of 0.4 is considered = -4dB.

Hence, CDMA has 100% frequency reuse. Use of same frequency in surrounding cells causes some additional interference.

In CDMA frequency, reuse efficiency is 0.67 (70% eff.) = -1.73dB

Advantages of CDMA

CDMA has a soft capacity. The greater the number of codes, the more the number of users. It has the following advantages:

- CDMA requires a tight power control, as it suffers from near-far effect. In other words, a user near the base station transmitting with the same power will drown the signal latter. All signals must have more or less equal power at the receiver.

- Rake receivers can be used to improve signal reception. Delayed versions of time (a chip or later) of the signal (multipath signals) can be collected and used to make decisions at the bit level.

- Flexible transfer may be used: Mobile base stations can switch without changing operator. Two base stations receive mobile signal and the mobile receives signals from the two base stations.

- Transmission Burst reduces interference.

Disadvantages of CDMA

The disadvantages of using CDMA are as follows:

- The code length must be carefully selected: A large code length can induce delay or may cause interference.

- Time synchronization is required.

- Gradual transfer increases the use of radio resources and may reduce capacity.

- As the sum of the power received and transmitted from a base station needs constant tight power control. This can result in several handovers.

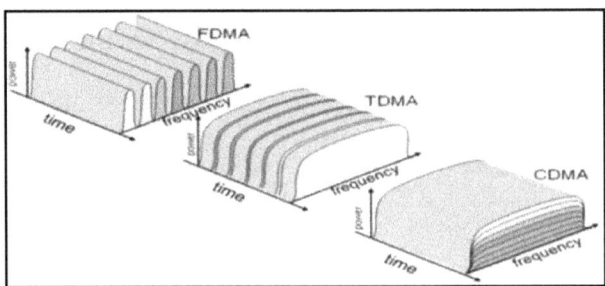

References

- Cheung, Nim K.; Nosu Kiyoshi; Winzer, Gerhard "Guest Editorial / Dense Wavelength Division Multiplexing Techniques for High Capacity and Multiple Access Communication Systems", IEEE Journal on Selected Areas in Communications, Vol. 8 No. 6, August 1990

- She, Alan; Capasso, Federico (17 May 2016). "Parallel Polarization State Generation". Scientific Reports. Nature. arXiv:1602.04463. Bibcode:2016NatSR...626019S. doi:10.1038/srep26019. Retrieved 27 June 2016

- What-is-multiplexer-and-types: efxkits.com, Retrieved 31 March 2018

- Ishio, H. Minowa, J. Nosu, K., "Review and status of wavelength-division-multiplexing technology and its application", Journal of Lightwave Technology, Volume: 2, Issue: 4, Aug 1984, p. 448–463

- Time-division-multiplexing-TDM: techtarget.com, Retrieved 11 May 2018

- Guowang Miao; Jens Zander; Ki Won Sung; Ben Slimane (2016). Fundamentals of Mobile Data Networks. Cambridge University Press. ISBN 1107143217

- Time-division-multiplexing, network-technologies: ecomputernotes.com, Retrieved 16 July 2018

Modulation and its Methods

The process of converting data into an electronic or optical carrier signal and its subsequent transmission is known as modulation. The two types of modulation discussed in this chapter are analog and digital modulation. This chapter's aim is to shed light on the topic of modulation and all of its related components for the benefit of the reader.

For a signal to be transmitted to a distance, without the effect of any external interference or noise addition and without getting faded away, it has to undergo a process called as Modulation. It improves the strength of the signal without disturbing the parameters of the original signal.

A message carrying a signal has to get transmitted over a distance and for it to establish a reliable communication; it needs to take the help of a high frequency signal which should not affect the original characteristics of the message signal.

The characteristics of the message signal, if changed, the message contained in it also alters. Hence, it is a must to take care of the message signal. A high frequency signal can travel up to a longer distance, without getting affected by external disturbances. We take the help of such high frequency signal which is called as a carrier signal to transmit our message signal. Such a process is simply called as Modulation. Modulation is the process of changing the parameters of the carrier signal, in accordance with the instantaneous values of the modulating signal.

Need for Modulation

Baseband signals are incompatible for direct transmission. For such a signal, to travel longer distances, its strength has to be increased by modulating with a high frequency carrier wave, which doesn't affect the parameters of the modulating signal.

Advantages of Modulation

The antenna used for transmission, had to be very large, if modulation was not introduced. The range of communication gets limited as the wave cannot travel a distance without getting distorted. Following are some of the advantages for implementing modulation in the communication systems:

- Reduction of antenna size.

- No signal mixing.

- Increased communication range.

- Multiplexing of signals.

- Possibility of bandwidth adjustments.

- Improved reception quality.

Signals in the Modulation Process

Following are the three types of signals in the modulation process:

- Message or Modulating Signal: The signal which contains a message to be transmitted is called as a message signal. It is a baseband signal, which has to undergo the process of modulation, to get transmitted. Hence, it is also called as the modulating signal.

- Carrier Signal: The high frequency signal, which has a certain amplitude, frequency and phase but contains no information, is called as a carrier signal. It is an empty signal and is used to carry the signal to the receiver after modulation.

- Modulated Signal: The resultant signal after the process of modulation is called as a modulated signal. This signal is a combination of modulating signal and carrier signal.

Types of Modulation

There are many types of modulations. Depending upon the modulation techniques used, they are classified as shown in the following figure:

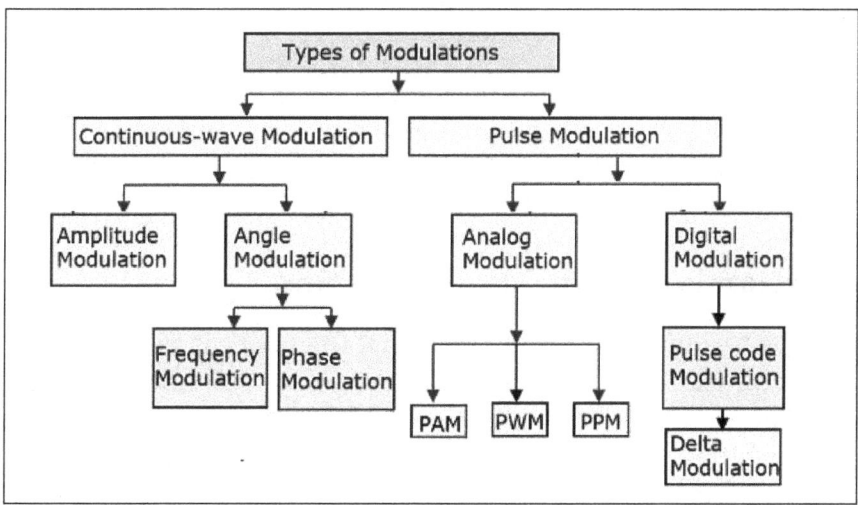

The types of modulations are broadly classified into continuous-wave modulation and pulse modulation.

Continuous-Wave Modulation

In continuous-wave modulation, a high frequency sine wave is used as a carrier wave. This is further divided into amplitude and angle modulation.

- If the amplitude of the high frequency carrier wave is varied in accordance with the instantaneous amplitude of the modulating signal, then such a technique is called as Amplitude Modulation.

- If the angle of the carrier wave is varied, in accordance with the instantaneous value of the modulating signal, then such a technique is called as Angle Modulation. Angle modulation is further divided into frequency modulation and phase modulation.

- If the frequency of the carrier wave is varied, in accordance with the instantaneous value of the modulating signal, then such a technique is called as Frequency Modulation.

- If the phase of the high frequency carrier wave is varied in accordance with the instantaneous value of the modulating signal, then such a technique is called as Phase Modulation.

Pulse Modulation

In Pulse modulation, a periodic sequence of rectangular pulses is used as a carrier wave. This is further divided into analog and digital modulation. In analog modulation technique, if the amplitude or duration or position of a pulse is varied in accordance with the instantaneous values of the baseband modulating signal, then such a technique is called as Pulse Amplitude Modulation (PAM) or Pulse Duration/Width Modulation (PDM/PWM), or Pulse Position Modulation (PPM).

In digital modulation, the modulation technique used is Pulse Code Modulation (PCM) where the analog signal is converted into digital form of 1s and 0s. As the resultant is a coded pulse train, this is called as PCM. This is further developed as Delta Modulation (DM).

Why Modulation is used in Communication

In the modulation technique, the message signal frequency is raised to a range so that it is more useful for transmission. The following points describe modulation's importance in the communication system. In signal transmission, the signals from various sources are transmitted through a common channel simultaneously by using multiplexers. If these signals are transmitted simultaneously with a certain bandwidth, they cause interference. To overcome this, speech signals are modulated to various carrier frequencies in order for the receiver to tune them to the desired bandwidth of his own choice within the range of transmission.

Another technical reason is antenna size; the antenna size is inversely proportional to the frequency of the radiated signal. The order of the antenna aperture size is at least one by a tenth of the wavelength of the signal. Its size is not practicable if the signal is 5 kHz; therefore, raising frequency by modulating process will certainly reduce the height of the antenna. Modulation is important to transfer the signals over large distances since it is not possible to send low-frequency signals for longer distances.

Similarly, modulation is also important to allocate more channels for users and to increase noise immunity:

- Modulating Signal: This signal is also termed as a message signal. It holds the data that has to be transmitted and so this termed as message signal. It is considered as the baseband signal where it undergoes a modulation process to get broadcasted or communicated. Because of this, it is the modulating signal.

- Carrier Signal: This is the high range of frequency signal which is with specific amplitude, frequency, and phase levels, but it does not hold any data. So, it is termed as carrier signal as it is an empty one. This is simply utilized to transmit the message to the receiver section after the process of modulation.

- Modulated Signal: The consequential signal that is obtained after the procedure of modulation is called a modulated signal. This is the product of both the carrier and modulating signals.

Analog Modulation

In this modulation, a continuously varying sine wave is used as a carrier wave that modulates the message signal or data signal. The Sinusoidal wave's general function is shown in the figure below, in which, three parameters can be altered to get modulation – they are mainly amplitude, frequency, and phase, so the types of analog modulation are:

- Amplitude modulation (AM).

- Frequency modulation (FM).

- Phase modulation (PM).

In amplitude modulation, the amplitude of the carrier wave is varied in proportion to the message signal, and the other factors like frequency and phase remain constant. The modulated signal is shown in the below figure, and its spectrum consists of a lower frequency band, upper-frequency band, and carrier frequency components. This type of modulation requires greater bandwidth, more power. Filtering is very difficult in this modulation.

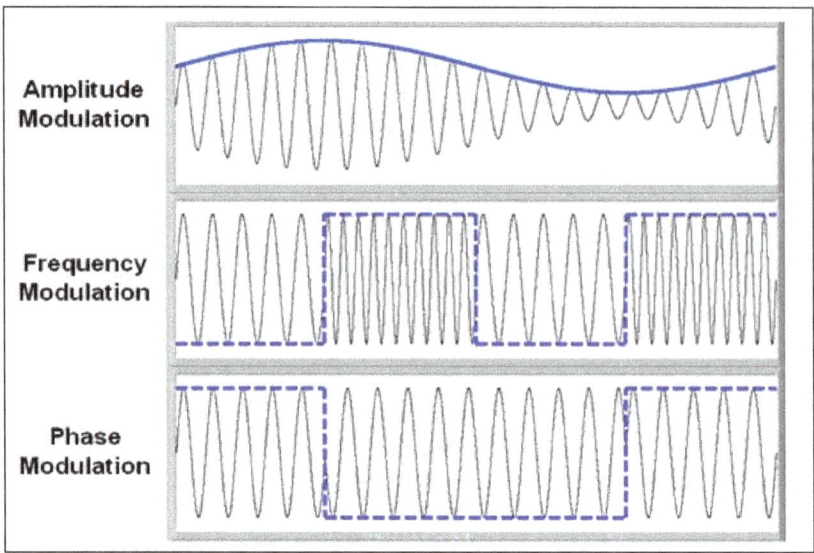

Traditional Analog Modulation Techniques

Modulation techniques are mainly used to transmit information in a given frequency band. The reason for that may be that the channel is band-limited, or that we are assigned a certain frequency band and frequencies outside that band is supposed to be used by others. Therefore, we are interested in the spectral properties of various modulation techniques.

The modulation techniques described here have a long history in radio applications. The information to be transmitted is normally an analog so called baseband signal. By that we understand a signal with the main part of its spectrum around zero. Especially, that means that the main part of the spectrum is below some frequency W, called the bandwidth of the signal.

We also consider methods to demodulate the modulated signals, i.e. to regain the original signal from the modulated one. Noise added by the channel will necessarily affect the demodulated signal. We separate the analysis of those demodulation methods into one part where we assume an ideal channel that does not add any noise and another part where we assume that the channel adds white Gaussian noise.

Amplitude Modulation

Amplitude modulation, normally abbreviated AM, was the first modulation technique. The first radio broadcasts were done using this technique. The reason for that is that AM signals can be detected very easily. Essentially, all you need is nonlinearity. Actually, almost any nonlinearity will suffice to detect AM signals. There have even been reports of people hearing some nearby radio station from their stainless steel kitchen sink. And some (including the author) have experienced that with a guitar amplifier.

The crystal receiver is a demodulator for AM that can be manufactured at a low cost, which helped making radio broadcasts popular. A convolution in the time domain corresponds to a multiplication in the frequency domain. In fact, the opposite is also true.

Theorem: Fourier transform of a multiplication.

Let a(t) and b(t) be signals with Fourier transforms A(f) and B(f). Then we have:

$$\mathcal{F}\{a(t)b(t)\}=(A*B)(f)$$

Proof: The proof is along the same line as the proof of Theorem 8, but starting with the inverse transform of the suggested spectrum. Based on our definitions, we have:

$$\mathcal{F}^{-1}\{(A*B)(f)\}=\int_{-\infty}^{\infty}(A*B)(f)e^{j2\pi ft}\,df=\int_{-\infty}^{\infty}\int_{-\infty}^{\infty}A(\phi)B(f-\phi)d\phi e^{j2\pi ft}df.$$

We can rewrite the expression above as:

$$\mathcal{F}^{-1}\{(A*B)(f)\}=\int_{-\infty}^{\infty}\int_{-\infty}^{\infty}A(\phi)B(f-\phi)e^{j2\pi ft}d\phi df.$$

Now, set $\lambda = f - \varphi$, and we get,

$$\mathcal{F}^{-1}\{(A*B)(f)\}=\int_{-\infty}^{\infty}\int_{-\infty}^{\infty}A(\phi)B(\lambda)e^{j2\pi(\lambda+\phi)t}d\tau d\lambda=\int_{-\infty}^{\infty}A(\phi)e^{j2\pi\phi t}d\phi\int_{-\infty}^{\infty}B(\lambda)e^{j2\pi\lambda t}\,d\lambda.$$

Finally, we identify the last two integrals as the inverse Fourier transforms of A(f) and B(f), and we get:

$$F^{-1}\{(A*B)(f)\}=a(t)b(t).$$

So, multiplying in the time domain corresponds to a convolution in the frequency domain.

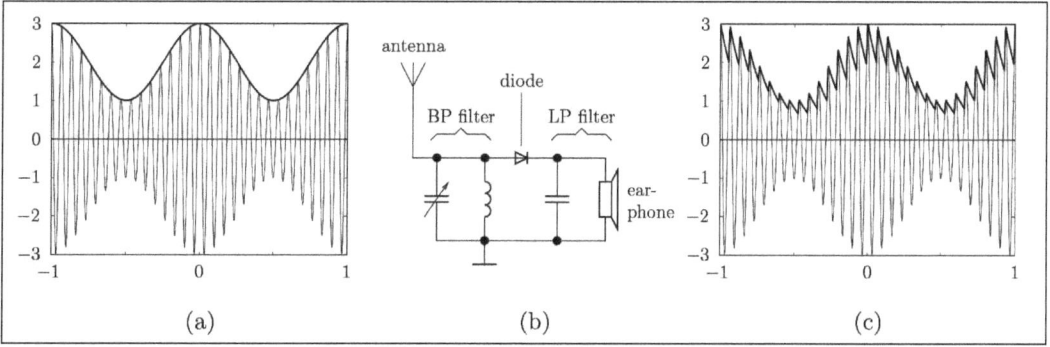

(a) A standard AM signal for the message m(t) = cos(2πt) with C = 2 and A = 1. The dark line is C + m(t). (b) Principle of an envelope detector. (c) The corresponding output from an envelope detector.

Standard AM

An AM signal, x(t), corresponding to the message signal, m(t), is given by the equation:

$$x(t) = A\left(c + m(t)\right)\cos\left(2\pi f_c t\right),$$

where f_c is referred to as the carrier frequency, A is some non-zero constant, and where the constant C is chosen such that $|m(t)| < C$ holds for all t. In Figure above standard AM signal is presented together with the message, which in this particular example is a cosine signal.

We mentioned that AM signals can be detected using a nonlinearity. The first AM receiver was the so called crystal receiver. It consists of an antenna, a resonance circuit (bandpass filter), a diode and a simple low-pass filter. It extracts the envelope C + m(t) from x(t), and is therefore often called an envelope detector. The diode in Figure is the nonlinearity that makes the detection possible. The few simple components make it possible to manufacture the receiver at a low cost. In addition to that, it doesn't even need a power source of its own. The power is taken directly from the antenna. The output power is of course very small, and only one listener could use the small earphone that was used. The AM signal is presented together with the output of an envelope detector. Note that the output is very similar to the original message. The mechanical parts in the earphone, and the ear will further low-pass filter the output, so the listener will hear almost the same signal as the one transmitted. Modern envelope detectors have amplifiers in various places and may be implemented digitally, but the basic construction is still a band pass filter, a diode (or some other rectifier) followed by a low-pass filter. An advantage of envelope detectors is that the BP filter that filters out everything except the intended frequency band is not critical. It is enough if its center frequency is approximately correct. In other words, it does not need to know the carrier frequency exactly, or the carrier phase for that matter.

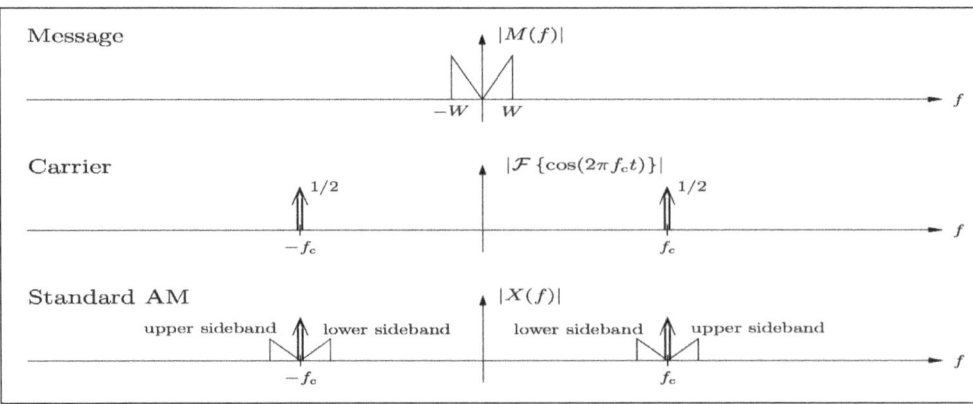

Spectrum for standard AM signals.

We wish to study the spectrum of AM signals. Since an AM signal is the product of a message and a carrier, that is easiest done based on Theorem. Thus, we need to find the Fourier transform of $\cos(2\pi f_c t)$. First consider,

$$\mathcal{F}^{-1}\{f - f_c\} = \int_{\infty}^{\infty} \delta(f - f_c)e^{j2\pi ft}df = e^{j2\pi f_c t},$$

where the first equality is given by the definition of the inverse Fourier transform, and where the last equality is given by the definition of the unit impulse. So, we have:

$$\mathcal{F}^{-1}\{\cos(2\pi f_c t)\} = \mathcal{F}\left\{\frac{e^{j2\pi f_c t} + e^{-j2\pi f_c t}}{2}\right\} = \frac{1}{2}(\delta(f - f_c) + \delta(f + f_c)).$$

Now we are ready to apply Theorem on x(t). Let M(f) be the spectrum of m(t) and let X(f) be the spectrum of x(t). Then we get:

$$X(f) = \mathcal{F}\{AC\cos(2\pi f_c t)\} + \mathcal{F}\{Am(t)\cos(2\pi f_c t)\}$$

$$= \frac{AC}{2}[\delta(f - f_c) + \delta(f + f_c)] + \frac{A}{2}[M(f - f_c) + M(f - f_c)].$$

It is left as an exercise to verify that the last equality holds. Here $\frac{AC}{2}$ ($\delta(f - f_c) + \delta(f + f_c)$) is referred to as the carrier, since that term corresponds to AC cos (2πfct). The other part of the spectrum, $\frac{A}{2}$ (M(f – f_c) + M(f + f_c)), is referred to as the sidebands. Those sidebands are the only parts of the spectrum that depend on the message m(t). The sidebands are called upper and lower sidebands based on where they are compared to the carrier frequency, according to the following:

- Upper sideband: $\frac{A}{2}$ (M(f – fc) + M(f + fc)) for |f| > fc.

- Lower sideband: $\frac{A}{2}$ (M(f – fc) + M(f + fc)) for |f| < fc.

Because of those two sidebands, this type of AM modulation is often called double sideband AM, abbreviated AM-DSB. There are also other versions of AM, but they cannot be detected using an envelope detector.

Suppressed Carrier Modulation

All the information about the message m(t) in standard AM is in the sidebands. The carrier itself does not carry any information, and in that respect the carrier corresponds to unnecessary power dissipation. One version of AM that cannot be detected using an envelope detector is called AM-SC or AM-DSB-SC, where SC should be interpreted as

Suppressed Carrier. For this type of modulation, the constant C is simply set to zero, i.e. we have:

$$x(t) = Am(t)\cos(2\pi f_c t),$$

and the corresponding spectrum is,

$$X(f) = \frac{A}{2}(M(f - f_c) + M(f + f_c)),$$

So the carrier is removed from the spectrum, as the name suggests. An AM-SC signal is presented together with the message and the corresponding envelope detector output, as well as the absolute value of the message. Note that the output from an envelope detector in this case is close to the absolute value of the message. Thus, an envelope detector cannot be used to receive AM-SC.

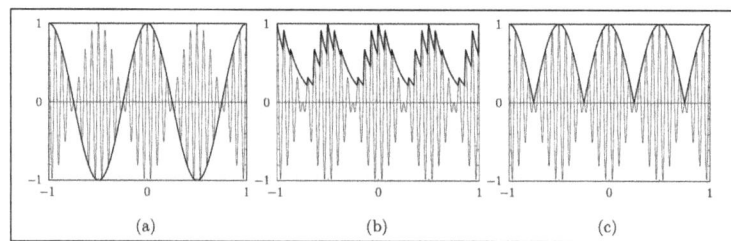

(a) An AM-SC signal for the message m(t) = cos(2πt) with A = 1. The thick line is m(t). (b) The corresponding output from an envelope detector. (c) The AM-SC signal together with |m(t)| for comparison.

Demodulation of AM-SC can instead be done by modulating once more. Let y(t) with spectrum Y (f) be the output of that modulation. Then we have, similarly as above,

$$y(t) = x(t)\cos(2\pi f_c t) = Am(t)\cos^2(2\pi f_c t) = \frac{A}{2}m(t)(1 + \cos(4\pi f_c t)),$$

and the corresponding spectrum is,

$$Y(f) = \frac{A}{2}(X(f - f_c) + X(f + f_c)) = \frac{A}{2}M(f) + \frac{A}{4}(M(f - 2f_c) + M(f + 2f_c)).$$

So, we have regained M(f), but we also have copies of M(f) centered around ±2f$_c$. The involved spectra are displayed. If W, the bandwidth of the message m(t), is smaller than f$_c$, which normally is the case, then those copies do not overlap with the original spectrum. Thus, we can use a suitable low-pass filter to remove the unwanted copies. The further away the unwanted copies are in the frequency domain, the simpler that filter can be. It should be noted that standard AM can also be demodulated using this method.

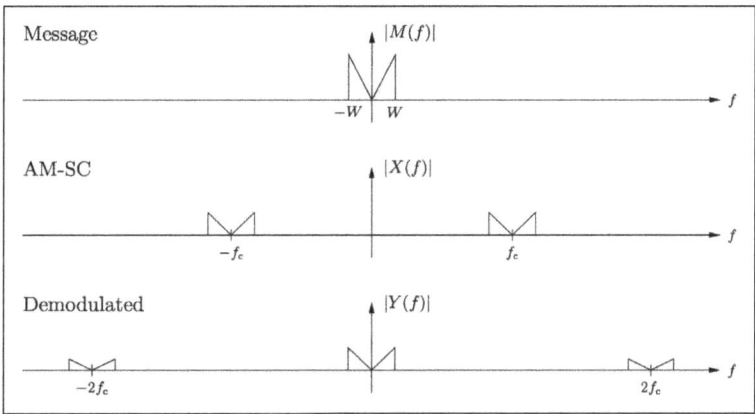

Modulation of AM-SC and demodulation by modulating again.

Single Sideband Modulation

Since there is a one-to-one relation between M(f) and M(−f), there is also a one-to-one relation between the two sidebands, at least if W is smaller than fc. So, in both standard AM and AM-SC, we actually transmit our data twice in the frequency domain. No information is lost if we only transmit one of the sidebands. This type of AM is referred to as SSB, which should be interpreted as Single SideBand. There are SSB versions of both standard AM and AM-SC, and they can be obtained by first generating standard AM or AM-SC, and then using a suitable band-pass filter to remove the unwanted sideband. Spectra of AM-SSB and AM-SSB-SC. Obviously, SSB modulation only needs half the bandwidth compared to original AM or AM-SC. SSB-modulated signals can also be demodulated by modulating again using AM-SC, and we still get copies near $\pm 2f_c$ that has to be removed by a low-pass filter. However these copies now contain only one sideband.

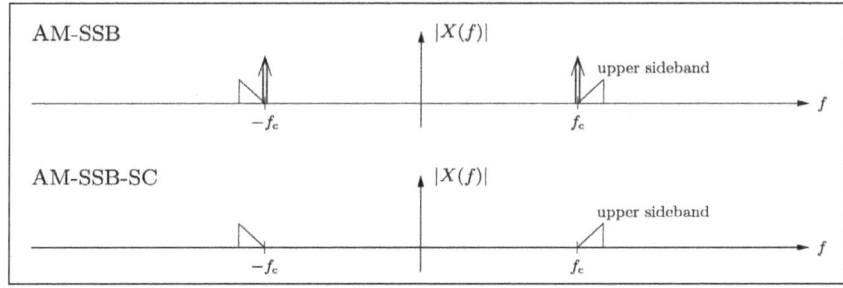

Spectra for AM-SSB and AM-SSB-SC.

Synchronization for AM Demodulation

Demodulation by re-modulation as described above is a method that can be used for detection of all variants of AM. However, that demands that we have a correct carrier available in the demodulator, with both correct frequency and at least approximately correct phase. For standard AM and for AM-SSB, where the carrier is available in the signal, it can easily be extracted from the received signal using a narrow BP-filter with center frequency f_c. For AM-SC, and for AM-SSB-SC, the absence of a carrier makes it impossible to extract the carrier in that way. One way for the receiver to extract a carrier signal from an AM-SC signal:

$$x(t) = Am(t)\cos(2\pi f_c t),$$

is to produce the square,

$$x^2(t) = A^2 m^2(t) \cos^2(2\pi f_c t) = \frac{A^2 m^2(t)}{2}\left(1 + \cos(4\pi f_c t)\right).$$

When we send information, the average of $m^2(t)$ is non-zero, which means that a scaled version of $\cos(4\pi f_c t)$ can be extracted from $x^2(t)$, again using a narrow BP-filter, but with center frequency 2fc. Extracting carriers in those ways will produce signals with a frequency that is the correct carrier frequency in the AM or AM-SSB case and twice the carrier frequency in the AM-SC case, but the amplitude can vary from time to time depending on the actual behaviour of the channel or the statistics of the information. Also, the extracted signal may include noise and parts of the sideband(s).

A clean carrier with both the correct frequency and well-defined amplitude, without any noise or residues from the sidebands, can be obtained from the extracted signal using a phase-locked loop (PLL). There are several variants of phase-locked loops in use, and a simple variant is displayed. The signals given in Figure assume that the input is already a clean sinusoid. In practice, the input is an extracted approximate carrier which, as noted above, is polluted with noise and residues from the sidebands. A phase-locked loop is a control loop that produces a sinusoid with constant amplitude, the correct frequency f_0 and an approximate phase φ. The voltage-controlled oscillator (VCO) is chosen such that the wanted carrier frequency corresponds to zero input. The frequency $f_{VCO}(Vin)$ of the output of the VCO is a function of the input voltage V_{in}, such that the derivative of that function is positive, i.e. we have $\dfrac{d}{dV_{in}} f_{VCO}(V_{in}) > 0$.

The carrier frequency in use may differ slightly from the wanted carrier frequency, and the phase-locked loop follows the carrier frequency, which means that the input to the VCO may differ slightly from zero. In Figure below that means that the signal produced by the PLL has phase φ for which $\cos(\varphi) \approx 0$ holds, in a point where $\cos(\varphi)$ has positive derivative. In other words, we have $\varphi \approx -\dfrac{\pi}{2} + k \cdot 2\pi$ for some integer k. We may assume any integer value of k, since the produced signal is the same for different values of k. So, we can for instance say that the signal has the phase $\varphi \approx -\dfrac{\pi}{2}$.

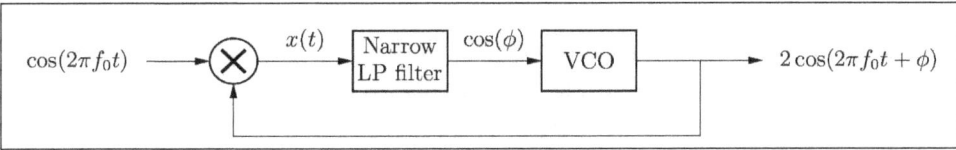

A phase-locked loop for generation of a well-defined carrier signal. The signal x(t) is given by $x(t) = \cos(2\pi f_0 t)\cos(2\pi f_0 t + \varphi) = \frac{1}{2}(\cos(4\pi f_0 t + \varphi) + \cos(\varphi))$. The device labelled VCO is a voltage controlled oscillator.

For AM-SC, where we have squared the signal, and where the frequency of the extracted signal is twice the carrier frequency, the feedback is equipped with a frequency doubler which can for instance be a squarer. The resulting output is then a sinusoid with the correct carrier frequency and phase approximately $-\frac{\pi}{4}$.

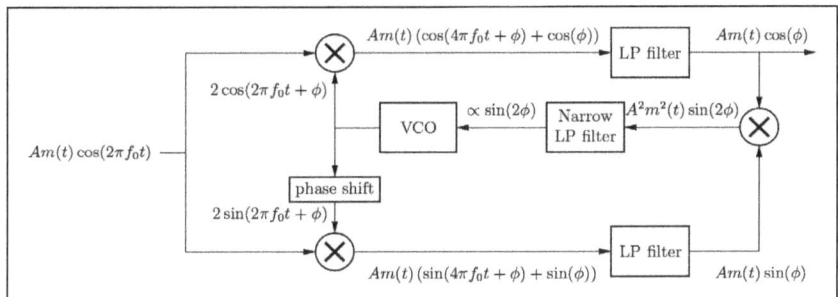

A Costas loop for detection of AM-SC.

A special type of phase-locked loop that is especially well suited for detection of AM-SC signals is the Costas loop. It extracts the carrier directly from the signal. Again, the VCO is chosen such that the wanted carrier frequency corresponds to zero input. Thus, the loop produces a sinusoid whose frequency is the carrier frequency with phase φ for which $\sin(2\varphi) \approx 0$ holds, in a point where $\sin(\varphi)$ has positive derivative. The resulting phase is therefore $\varphi \approx 0$. The output of the Costas loop is the message m(t) scaled by $\cos(\varphi)$, but since we have $\varphi \approx 0$, we also have $\cos(\varphi) \approx 1$.

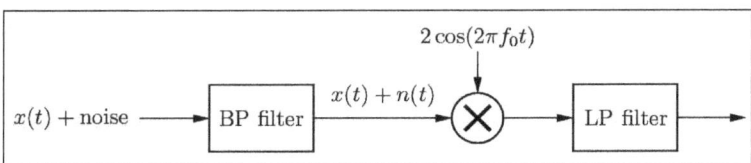

Demodulation of AM signals in the presence of noise.

The reason that the Costas loop works is the presence of both sidebands, and the one-tone mapping between the two sidebands. The two sidebands point at the carrier frequency and the phase information in the sidebands gives us the carrier phase. The Costas loop can therefore not be used for AM-SSB-SC. There is simply no way to extract the carrier from an AM-SSB-SC signal, due to the fact that the carrier is not available in the signal, and nothing in the spectrum gives any hint about the carrier frequency. That is the price we have to pay for suppressing both the carrier and one of the sidebands. Therefore, there are a number of modifications of AM-SSB-SC that makes it possible to extract a carrier anyway. The most simple method is not to suppress the carrier completely. Then the – however weak – carrier can be extracted from the signal. Another method is to send a short carrier burst, and let a PLL lock on to that burst. After the burst, the oscillator continues producing an internal carrier based on that burst. Of course, the oscillator will most probably diverge from the used carrier eventually. Therefore the carrier burst is repeated regularly. A third possibility is to keep a small part of the removed sideband, and in the receiver filter the other sideband similarly, and then extract a carrier using one of the methods above.

Impact of Noise in AM Demodulation

We would like to analyze the impact of noise on demodulation of AM signals. For this analysis we need to make some assumptions about the noise and about the demodulation. The first assumption is that the noise is dominated by thermal noise, and that it is independent of the message. We assume that the received signal is filtered by an ideal BP filter that exactly matches the bandwidth of the AM signal before demodulation. We assume that the demodulation is done by remodulation by $2 \cdot \cos(2\pi f_0 t)$ as in the Costas loop. We also assume that the demodulated signal is filtered by an ideal LP filter that exactly matches the bandwidth of the message.

We need to introduce some notation. Let W denote the bandwidth of the message. Let N_0 denote the one-sided power spectral density of the assumed white Gaussian noise, and let n(t) denote the noise after the BP filter. Also, introduce the following notation for the involved powers.

- P: The (expected) power of the message m(t).

- P_{m-mod}: The (expected) power of the received modulated signal x(t). Note that this means that A includes impacts of the channel.

- P_m: The (expected) power of the message after demodulation and LP filter.

- P_{n-mod}: The expected power of the ideally BP-filtered noise n(t) before demodulation.

- P_n: The expected power of the demodulated and LP-filtered noise.

We define the signal-to-noise ratio P_m/P_n after demodulation. We will compare this signal-to-noise ratio for DSB and SSB modulation using the same sent power P_{m-mod} transmitted over a channel with the same N_0.

First we consider AM-SC. Then we have the signal,

$$x(t) = Am(t)\cos(2\pi f_c t),$$

with bandwidth 2W and expected power $P_{m-mod} = A^2P/2$ since the carrier $\cos(2\pi f_c t)$ has average power 1/2. After demodulation, we regain Am(t), which means that we have Pm = A^2P. For the noise, we have $P_{n-mod} = 2WN_0$. The demodulated noise:

$$n(t). 2\cos(2\pi f_c t)$$

has expected power $2P_{n-mod}$, since the carrier $2\cos(2\pi f_c t)$ has average power 2. Half of that expected power is in the frequency interval $|f| < W$, while the other half is in the frequency interval $2f_0 - W < |f| < 2f_0 + W$. The latter part is removed by the LP filter, leaving us with $P_n = P_{n-mod}$. Finally, that gives us the signal-to-noise ratio:

$$\frac{P_m}{P_n} = \frac{A^2P}{2WN_0}$$

For AM-SSB-SC, one of the sidebands from AM-SC is removed, which means that the power Pm−mod is reduced to half that of AM-SC. To produce an SSB signal with the same power as in the DSB case, we therefore need to amplify the signal by $\sqrt{2}$. So, we start with,

$$x(t) = \sqrt{2}\, Am(t)\cos(2\pi f_c t),$$

and filter out one of the sidebands. Then we have the same sent power Pm−mod = $A^2P/2$. After demodulation, we get a scaled version of the message. More precisely, the output is $\frac{A}{\sqrt{2}} m(t)$, which has power $P_m = A^2P/2$. For the noise, we have $P_{n-mod} = WN_0$, since the bandwidth is W. The demodulated noise:

$$n(t) \cdot 2\cos(2\pi f_c t)$$

still has expected power $2P_{n-mod}$, since the carrier $2\cos(2\pi f_c t)$ has average power 2. Half of that expected power is in the frequency interval $|f| < W$, while the other half is in the frequency interval $2f_0 - W < |f| < 2f_0 + W$. Actually, the other half of the power is in the interval $2f_0 - W < |f| < 2f_0$ if the lower sideband is used, or in the interval $2f_0 < |f| < 2f_0 + W$ if the upper sideband is used. In any case, the part of the spectrum that is near $2f_0$ is removed by the LP filter, leaving us with $P_n = P_{n-mod}$. Finally, that gives us the signal-to-noise ratio,

$$\frac{P_m}{P_n} = \frac{A^2P}{2WN_0},$$

i.e. the same signal-to-noise ratio as for DSB.

Digital Modulation

Digital modulation is the process of encoding a digital information signal into the amplitude, phase, or frequency of the transmitted signal. The encoding process affects the bandwidth of the transmitted signal and its robustness to channel impairments. In general, a modulation technique encodes several bits into one symbol, and the rate of symbol transmission determines the bandwidth of the transmitted signal. Since the signal bandwidth is determined by the symbol rate, having a large number of bits per symbol generally yields a higher data rate for a given signal bandwidth. However, the larger the number of bits per symbol, the greater the required received SNR for a given target BER.

Digital modulation techniques may be linear or nonlinear. In linear modulation the amplitude and/or phase of the transmitted signal varies linearly with the digital modulating signal, whereas the transmitted signal amplitude is constant for nonlinear techniques.

Linear modulation techniques, including all forms of quadrature-amplitude modulation (QAM) and phase-shift-keying (PSK), use less bandwidth than nonlinear techniques, including various forms of frequency/minimum-shift-keying (FSK and MSK). Since linear techniques encode information into the amplitude and phase of linear modulation, this type of modulation is more susceptible to amplitude and phase fluctuations caused by multipath flat-fading. In addition, the amplifiers used for linear modulation must be linear, and these amplifiers are more expensive and less efficient than nonlinear amplifiers. Thus, the bandwidth efficiency of linear modulation is generally obtained at the expense of hardware cost, power, and higher BERs in fading. Linear modulation techniques are used in most wireless LAN products, whereas nonlinear techniques are used in most cellular and wide area wireless data systems.

Linear modulation techniques can be detected coherently or differentially. Coherent detection requires the receiver to obtain a coherent phase reference for the transmitted signal. This is difficult to do in a rapidly fading environment, and also increases the complexity of the receiver. Differential detection uses the previously detected symbol as a phase reference for the current symbol. Because this detected symbol is a noisy reference, differential detection requires roughly twice the power of coherent detection for the same BER. Moreover, if the channel is changing rapidly, then differential detection is not very accurate, since the channel phase may change considerably over one symbol time. As a result, rapidly changing channels with differential detection have an irreducible error floor, that is, the BER of the channel has a lower bound (error floor) that cannot be reduced by increasing the received SNR. This error floor increases as the rate of channel variation (the channel Doppler) increases and decreases as the data rate increases (since a higher data rate corresponds to a shorter bit time, so the channel phase has less time to decorrelate between bits). For high-speed wireless data (above 1 Mbps), the error floor is quite low at user speeds below 60 mph, but at lower data rates the error floor becomes significant, thereby preventing the use of differential detection.

Phase Shift Keying

Both OOK and FSK were shown to result in a probability of error of:

$$P_e = \text{erfc}\sqrt{\frac{E_b}{\eta}},$$

where Eb is the average energy per bit.

Stremler's discussion of why OOK and FSK don't perform. In the simplest case a binary phase-shift keyed (BPSK) signal takes the form:

$$s(t) = m(t)\cos(\omega_c t),$$

where ωc is the carrier frequency, and $m(t)$ is a polar binary baseband signal taking on the value 1 for a mark and −1 for a space. Because there are only two different signals, and they differ only by a change of sign, this signalling scheme is also called phase-reversal keying (PRK). In this case the signals sent are,

$$s_1(t) = A\cos(\omega_1 t), \qquad 0 < t \leq T.$$

$$s_2(t) = -A\cos(\omega_1 t), \qquad 0 < t \leq T.$$

When the baseband signal consists of square pulses with amplitude taking on values of +1 and −1 with equal probability, it can be shown that the power spectral density is:

$$P(f) = \frac{1}{4}\left[\frac{\pi T(f - f_c)}{\pi T(f - f_c)}\right]^2 + \frac{1}{4}\left[\frac{\pi T(f + f_c)}{\pi T(f + f_c)}\right]^2.$$

Here the data rate is R = 1/T bits/sec. This is essentially the same as the PSD obtained by assuming a deterministic alternating square wave taking on values of +1 and −1 in sequence, which can be considered a worst case. Coherent or synchronous detection of a PRK signal can be performed by a system of the following form:

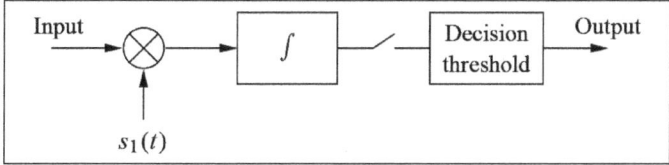

Because the signals differ only in sign, there is no need to have two matched filters in the detection. Instead the matched filter nominally takes on a value of E when $s_1(t)$ is present at the input, and −E when $s_2(t)$ is present. The noise variance at the output of the filter is $E\eta/2$, so the overall probability of error at the receiver is:

$$P_e = \text{erfc}\left(\frac{2E}{\sqrt{E\eta/2}}\right) = \text{erfc}\sqrt{\frac{2E}{\eta}}.$$

Since the signals have the same energy when a zero and a one are transmitted, this can be written as:

$$P_e = \text{erfc}\sqrt{\frac{2E_b}{\eta}}.$$

For the same communication channel, PRK therefore requires half the transmission energy for the same bit error probability as OOK and FSK. The bandwidth required is also quite modest when compared with these other signalling schemes.

The PRK signal can also be obtained in the context of an ASK system modulated by a polar baseband signal. However, there is a 3dB advantage in using PRK over OOK. A coherent reference for synchronous detection cannot be obtained by the use of an ordinary phase-locked tracking loop, since there are no spectral line components at $\pm f_c$. However, since the signal has a spectrum that is symmetric with respect to the (suppressed) carrier frequency, either a squaring loop or a Costas PLL can be used to obtain synchronization. The diagram for a squaring loop in a coherent detector is shown below:

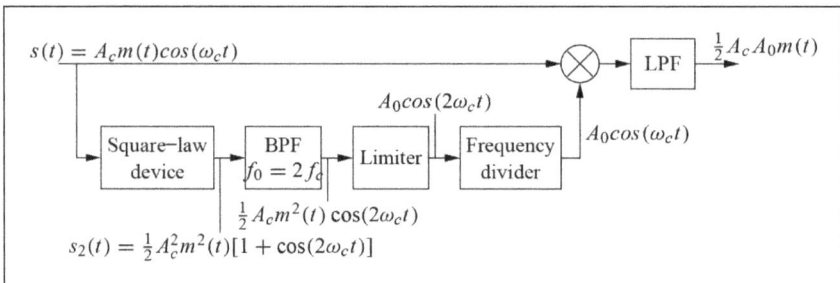

Alternatively, a Costas phase-locked loop can also perform the task:

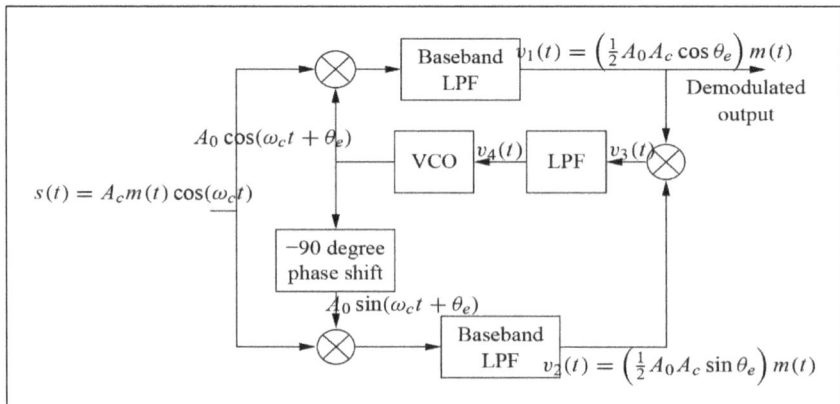

This can be analysed by assuming that the VCO is locked to the input suppressed carrier frequency f_c, with a constant phase error of θ_e. Then the voltages v1(t) and v2(t) are obtained at the output of the baseband low-pass filters as shown. Since θ_e is small, the amplitude of $v_1(t)$ is relatively large compared to that of $v_2(t)$. Furthermore, $v_1(t)$ is proportional to m(t), so it is the demodulated output. The product voltage $v_3(t)$ is:

$$v_3(t) = 1/2(1/2A_0A_C)^2 m^2(t)\sin 2\theta_e.$$

The voltage $v_3(t)$ is filtered with a LPF that has cutoff frequency near DC so, that this filter acts as an integrator to produce the DC VCO control voltage:

$$v_4(t) = K\sin 2\theta_e$$

where K = $1/2(1/2A_0A_c)^2$\{m2(t)\} and \{m2(t)\} is the DC level of $m^2(t)$. This DC control voltage is sufficient to keep the VCO locked to f_c with a small phase error θ_e.

Both of these solutions have one disadvantage — a 180 degree phase ambiguity. It can be shown that the noise performance of the squaring loop and the Costas PLL are equivalent, so the choice of which to implement depends on the relative cost of the loop components and the accuracy that can be realized when each component is built.

Phase-shift keyed signals cannot be detected incoherently. However, a partially coherent technique can be used whereby the phase reference for the present signalling interval is provided by a delayed version of the signal that occurred during the previous sampling interval. A differential PSK decoder takes the following form:

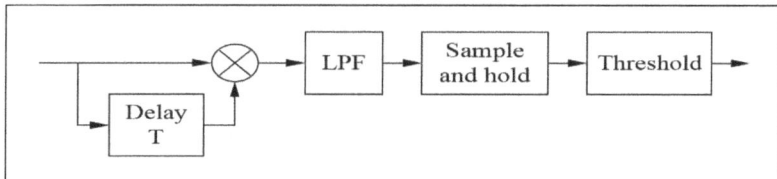

If a zero-noise BPSK signal is applied to the receiver input, the output of the sample-and-hold circuit will be positive (binary 1) if the present data bit and the previous data bit are the same; the output is negative (binary zero) if the two data bits are different. Thus if the data in the BPSK is differentially encoded, then the decoded sequence will be recovered at the output of this receiver. A more general representation for a PSK signal takes the following form:

$$s(t) = A\sin(\omega_c t + \Delta\theta m(t)).$$

Assume that m(t) has peak values of ±1. This expression can be expanded as:

$$s(t) = mA\sin\omega_c t + m(t)\sqrt{1 - m^2}A\cos\omega_c t.$$

The first term contains the pilot carrier, while the second carries the data. The average power in the carrier is $m^2 A^2/2$, and the power in the modulation component is $(1 - m^2)$ $A^2/2$. Thus a fraction m^2 of the total power in the modulated waveform is allocated to the carrier. It follows that the carrier component is zero in a PRK waveform, for which $\Delta\theta = \pi/2$. Stremler indicates that the probability of error for BPSK is,

$$P_e = \mathrm{erfc}\sqrt{2E(1-m^2)/\eta}.$$

Thus the effect of allocating a fraction m2 of the total power to the carrier is to degrade P_e by an equivalent S/N loss of 10 $\log_{10}(1 - m^2)$ dB. However, the resulting waveform has a spectral line at the carrier, which can be found using a conventional phase-locked loop.

Frequency Shift Keying

In frequency-shift keying, the signals transmitted for marks (binary ones) and spaces (binary zeros) are:

$$s_1(t) = A\cos(\omega_1 t + \theta_c), \qquad 0 < t \le T.$$

$$s_2(t) = A\cos(\omega_2 t + \theta_c), \qquad 0 < t \le T.$$

respectively. This is called a discontinuous phase FSK system, because the phase of the signal is discontinuous at the switching times. A signal of this form can be generated by the following system:

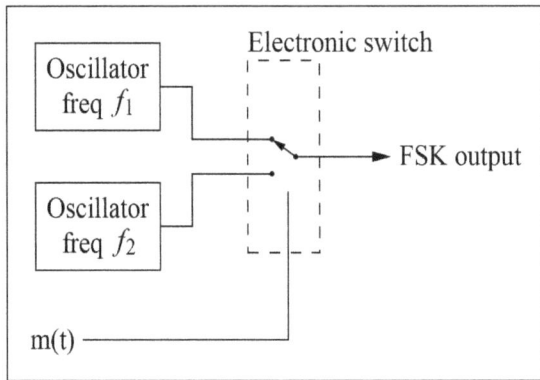

If the bit intervals and the phases of the signals can be determined (usually by the use of a phase-lock loop), then the signal can be decoded by two separate matched filters:

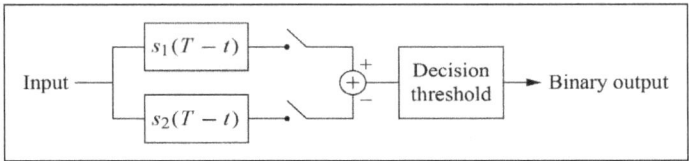

The first filter is matched to the signal $s_1(t)$, and the second to $s_2(t)$. Under the assumption that the signals are mutually orthogonal, the output of one of the matched filters will be E and the other zero (where E is the energy of the signal). Decoding of the band pass signal can therefore be achieved by subtracting the outputs of the two filters, and comparing the result to a threshold.

If the signal $s_1(t)$ is present then the resulting output will be $+E$, and if $s_2(t)$ is present it will be $-E$. Since the noise variance at each filter output is $E\eta/2$, the noise in the difference signal will be doubled, namely $\sigma^2 = E\eta$. Since the overall output variation is $2E$, the probability of error is:

$$P_\epsilon = \mathrm{erfc}\left(\frac{2E}{2\sqrt{E n}}\right) = \mathrm{erfc}\sqrt{\frac{E}{\eta}}.$$

The overall performance of a matched filter receiver in this case is therefore the same as for ASK. The frequency spectrum of an FSK signal is difficult to obtain — this is a general characteristic of FM signals. However, some rules of thumb can be developed. Consider the case where the binary message consists of an alternating sequence of zeros and ones. If the two frequencies are each multiples of $1/T$ (e.g. $f_1 = m/T$ and $f_2 = n/T$) and are synchronized in phase, then the FSK wave is a periodic function:

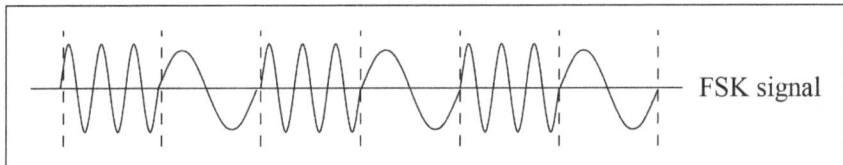

FSK signal

This can be viewed as the linear superposition of two OOK signals, one delayed by T seconds with respect to the other. Since the spectrum of an OOK signal is:

$$F_c(\omega) = \int_{\inf ty}^{\infty} Am(t)\cos(\omega_c t)e^{-j\omega t}dt = \int_{-\infty}^{\infty} A/2m(t)[e^{j\omega_c t} + e^{-j\omega_c t}]dt$$
$$= A/2[M(\omega - \omega_c) + M(\omega + \omega_c)],$$

where $M(\omega)$ is the transform of the baseband signal m(t), the spectrum of the FSK signal is the superposition of two of these spectra, one for $\omega_1 = \omega_c - \Delta\omega$ and the other for $\omega_2 = \omega c + \Delta\omega$. An example is shown below for positive frequencies under the assumption that $2\Delta f\, T \gg 1$:

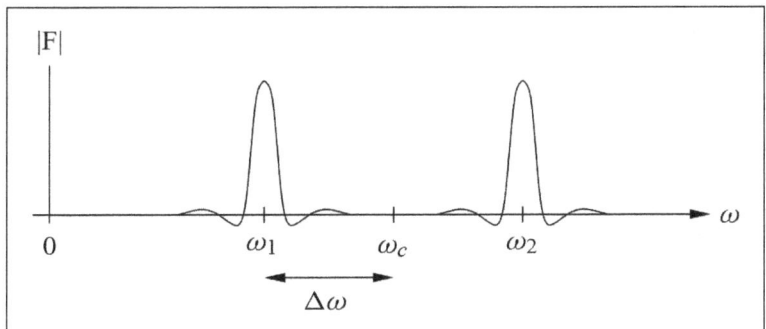

The bandwidth of the periodic FSK signal is then $2\Delta f + 2B$, with B the bandwidth of the baseband signal. Nonsynchronous or envelope detection can be performed for FSK signals. In this case the receiver takes the following form:

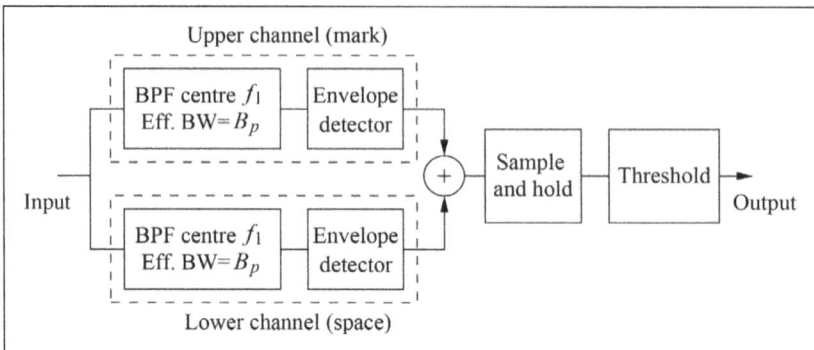

The bit error probability can be shown to be:

$$P_\varepsilon = \frac{1}{2}e^{-E/2\eta},$$

which under normal operating conditions corresponds to less than a 1dB penalty over coherent detection. In practice almost all FSK receivers are of this form.

In order for envelope detection to be successful, the peaks in the frequency domain at $\omega_c - \Delta\omega$ and $\omega_c + \Delta\omega$ must be widely separated with respect to the bandwidth of the baseband signal. This requires $2\Delta f\, T > 1$.

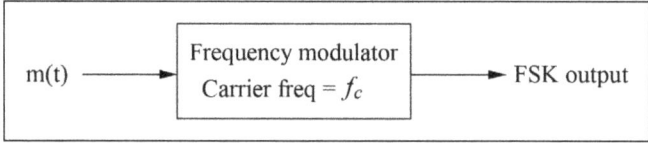

In this case the FSK signal can be represented by:

$$s(t) = A\cos\left[\omega_c t + D_f \int_{-\infty}^{t} m(\lambda)d\lambda\right].$$

Overly sharp transitions in the phase of the output signal can be restricted by band-limiting the input to the VCO. Note that FSK is not true frequency modulation, and does not provide the wide-band noise reduction properties associated with FM.

Example: A rectangular-pulse polar baseband signal is used to modulate an RF carrier in FSK. If the baseband signal has a data rate of 200 kbit/sec and the two RF frequencies are 150 kHz apart, determine the bandwidth.

The bit period in the baseband signal is $T = 1/200000$ seconds, and the baseband pulses are rectangular. The bandwidth of the baseband signal (to the first null) is given by

$B = 1/(2T)$. For the RF components, $2\Delta f = 150$ kHz. The bandwidth of the FSK signal is therefore:

$$2\Delta f + 2B = 150\text{khz} + 200 \text{ kHz} = 350 \text{ kHz.}$$

Comparison of Digital Modulation Schemes

The simple band-pass signaling schemes discussed so far have relative strengths and weaknesses.

Amplitude-Shift Keying

The net probability of error for a coherently detected OOK system is:

$$P_\varepsilon = \text{erfc}\sqrt{\frac{E}{2\eta}},$$

where E is the bit energy on transmission of a mark. This expression is relevant if the peak power is the important design parameter. If marks and spaces occur with equal probability, then this can be written as:

$$P_\varepsilon = \text{erfc}\sqrt{\frac{E_b}{\eta}},$$

with E_b the average bit energy. Because of the presence of a large carrier component, non-coherent detection of OOK is also possible, so simple envelope detection can be used in the receiver. Synchronous detection offers only about a 1dB improvement over envelope detection.

The PSD of ASK is centered at ωc, and has an identical shape to the corresponding on-off keyed baseband signal. Since the bandwidth has been doubled by the modulation, the theoretical maximum bandwidth efficiency is 1bps/Hz.

Transmitters for ASK are easy to build, as are non-coherent receivers. OOK systems are often used in short-range miniature telemetry. The decision threshold in the receiver does however have to be adjusted with changes in received signal levels, usually by means of an automatic gain control circuit.

Frequency-Shift Keying

The probability of error for coherent FSK is:

$$P_\varepsilon = \text{erfc}\sqrt{\frac{E}{\eta}}.$$

Since the signal is active all the time, this can be expressed in terms of the average bit energy E_b as:

$$P_\varepsilon = \text{erfc}\sqrt{\frac{E_b}{\eta}}.$$

In terms of average power required, the performance of FSK is therefore the same as for ASK. However, in terms of peak power, FSK has a 3dB advantage over ASK.

FSK systems operate symmetrically about a zero decision-threshold regardless of the carrier signal strength, so threshold adjustments need not be made. Additionally, there is little difference in complexity in FSK transmitters over ASK. Receiver complexity may vary, however, according to whether coherent or non-coherent detection is used.

Non-coherent detection of FSK is quite simple to perform, and is popular for low-to-medium data transmission rates. However, the frequencies used must then satisfy the condition $2\Delta f\, T \gg 1$, so that the peaks in the PSD are well-separated. The bandwidth required in this case is $2\Delta f + 2B$, where B is the baseband bandwidth.

For coherent detection Δf can be made as small as desired, but cases for $2\Delta f\, T < \frac{1}{2}$ result in a S/N penalty. Bandwidths for FSK transmission intended for coherent demodulation are typically equal to or slightly greater than those used for ASK.

Phase-Shift Keying

The error probability for PSK is:

$$P_\varepsilon = \text{erfc}\sqrt{\frac{2E}{\eta}},$$

Or

$$P_\varepsilon = \text{erfc}\sqrt{\frac{2E_b}{\eta}}.$$

Thus, PSK systems require less transmitted power for a given probability of error than ASK or FSK systems.

Synchronous detection of PSK signals is required, due to the absence of a large carrier component. Carrier recovery is therefore more complex and expensive. DPSK systems are often a good compromise, offering simpler circuitry at a small performance cost. The PSD of a PSK waveform is centered around ω_c, and has an identical shape to that of the double sideband modulating spectral density.

For PSK with $\Delta\theta < \pi/2$, there is a carrier component and the PSD has an impulse at the carrier frequency. The carrier component need not be large with respect to the sidebands. The theoretical bandwidth efficiency of PSK systems is 1bps/Hz.

Permissions

All chapters in this book are published with permission under the Creative Commons Attribution Share Alike License or equivalent. Every chapter published in this book has been scrutinized by our experts. Their significance has been extensively debated. The topics covered herein carry significant information for a comprehensive understanding. They may even be implemented as practical applications or may be referred to as a beginning point for further studies.

We would like to thank the editorial team for lending their expertise to make the book truly unique. They have played a crucial role in the development of this book. Without their invaluable contributions this book wouldn't have been possible. They have made vital efforts to compile up to date information on the varied aspects of this subject to make this book a valuable addition to the collection of many professionals and students.

This book was conceptualized with the vision of imparting up-to-date and integrated information in this field. To ensure the same, a matchless editorial board was set up. Every individual on the board went through rigorous rounds of assessment to prove their worth. After which they invested a large part of their time researching and compiling the most relevant data for our readers.

The editorial board has been involved in producing this book since its inception. They have spent rigorous hours researching and exploring the diverse topics which have resulted in the successful publishing of this book. They have passed on their knowledge of decades through this book. To expedite this challenging task, the publisher supported the team at every step. A small team of assistant editors was also appointed to further simplify the editing procedure and attain best results for the readers.

Apart from the editorial board, the designing team has also invested a significant amount of their time in understanding the subject and creating the most relevant covers. They scrutinized every image to scout for the most suitable representation of the subject and create an appropriate cover for the book.

The publishing team has been an ardent support to the editorial, designing and production team. Their endless efforts to recruit the best for this project, has resulted in the accomplishment of this book. They are a veteran in the field of academics and their pool of knowledge is as vast as their experience in printing. Their expertise and guidance has proved useful at every step. Their uncompromising quality standards have made this book an exceptional effort. Their encouragement from time to time has been an inspiration for everyone.

The publisher and the editorial board hope that this book will prove to be a valuable piece of knowledge for students, practitioners and scholars across the globe.

Index

Lightning Source UK Ltd.
Milton Keynes UK
UKHW052051080922
408583UK00002B/21